Ralph Tate

A Handbook of the Flora of Extratropical South Australia

Containing the flowering plants and ferns

Ralph Tate

A Handbook of the Flora of Extratropical South Australia
Containing the flowering plants and ferns

ISBN/EAN: 9783337315351

Printed in Europe, USA, Canada, Australia, Japan

Cover: Foto ©Andreas Hilbeck / pixelio.de

More available books at **www.hansebooks.com**

A HANDBOOK

OF THE

FLORA

OF

EXTRATROPICAL SOUTH AUSTRALIA

CONTAINING THE

FLOWERING PLANTS AND FERNS.

BY

RALPH TATE, F.L.S., F.G.S.,

HONORARY MEMBER OF THE ROYAL SOCIETY, NEW SOUTH WALES; NATURAL HISTORY
PHILOSOPHICAL SOCIETIES OF BELFAST AND WHITBY; FIELD NATURALISTS' CLUB OF BELFAST
AND MELBOURNE. CORRESPONDING MEMBER OF THE ACADEMY OF SCIENCES,
PHILADELPHIA; LINNEAN SOCIETY, NEW SOUTH WALES; ROYAL
SOCIETY, TASMANIA. PROFESSOR OF NATURAL HISTORY
IN THE UNIVERSITY OF ADELAIDE.

ADELAIDE:
PUBLISHED BY THE EDUCATION DEPARTMENT.

1890.

CONTENTS.

- Key to the System of South Australian Plants.

- A Classified List of the Native Species with Annotations indicating their Distribution within the Province. With a map.

- Explanation of Specific Names.

- Index of the Orders and Genera, with Explanation of the Generic Names.

PREFACE.

This work is intended for those who have mastered the elements of botany and who wish to be acquainted, as rapidly and readily as may be, with the name and systematic position of any our of Native Plants. It is purposely kept brief, and, though too abridged to serve as a sole source of information, yet it is issued to meet the need of a handy work of reference, since the Flora Australiensis is too bulky and too expensive.

The plan of the Key is adopted chiefly from the Flora Australiensis, and a little practice will suffice to enable the tyro to make use of it, especially if he select at first a few known species. "The student having a plant to determine, will first take the general table of Natural Orders, and examining his plant at each step to see which alternative agrees with it, will be led on to the Order to which it belongs. If it agrees, he will follow the same course with the table of the genera of that Order, and again with the table of species of the genus. But in each case, if he finds that his plant does not agree with the description of the genus or species to which he has been referred, he must revert to the beginning and carefully go through every step of the investigation before he can be satisfied. A fresh examination of his specimen, or of others of the same plant, a critical consideration of the meaning of every expression in the characters given, may lead him to detect some minute point overlooked or mistaken, and put him into the right way. Species vary within limits which is very difficult to express in words, and it proves often impossible, in framing these analytical tables, so to divide the genera and species, that those which come under one alternative should absolutely exclude the others; in such doubtful cases both alternatives must be tried." *Bentham.* Special attention is directed to the characters printed in italics.

The determination of the systematic position of a plant is often difficult and at times impossible without the aid of matured fruits.

The specific characters are in most cases comparative only as regards South Australian species.

The definitions of the generic and specific names will it is thought be of some aid in associating the name with a botanical character, though not always happily selected.

It will always afford me pleasure to assist any correspondent with his difficulties, and to receive authentic specimens of species unrecorded for a district.

Students using this book are recommended to insert in their proper places the corrections, &c., p. 301-3.

My thanks are due to the Hon. the Minister of Education, under whose departmental auspices this volume has been published; and to a colleague for revising parts III. and IV.

University of Adelaide,
    October, 1889.

FLORA

OF

EXTRATROPICAL SOUTH AUSTRALIA.

Key to the System of South Australian Plants.

CLASSES AND MAIN DIVISIONS OF PLANTS.

———o———

CLASS I.—DICOTYLEDONS.

Embryo with two, rarely more, seed-lobes. Floral organs usually in fours or fives. Veins of leaves mostly reticulate. Woody stems with a central pith, surrounded by concentric layers of wood, and an exterior bark.

DIVISION I.—ANGIOSPERMS.

Stigma present. Ovule within an ovary. Cotyledons two, rarely more.

Sub-Class I.—Choripetaleae Hypogynae.

Petals distinct, rarely united, sometimes absent. Ovary superior, quite free from the other floral structures. Stamens inserted at the base or below the ovary.

Sub-Class II.—Choripetaleae Perigynae.

Petals usually distinct, rarely absent. Ovary free from or adnate to the calyx. Stamens inserted on the calyx.

Sub-Class III.—Synpetaleae Perigynae.

Petals united, rarely distinct or absent. Ovary adnate to the calyx tube. Stamens inserted on the corolla, or in Candolleaceae adnate to the style.

Sub-Class IV.—Synpetaleae Hypogynae.

Petals united, rarely distinct or absent. Ovary superior, free from the calyx. Stamens inserted on the corolla, or rarely at the base of the ovary.

DIVISION II.—GYMNOSPERMS.

Flowers strictly unisexual, without calyx or corolla. Stigma absent; ovules naked, in the axils of scales forming a cone. Cotyledons, two or more.

CLASS II.—MONOCOTYLEDONS.

Embryo with one seed-lobe. Floral organs usually in threes. Veins of leaves mostly parallel and longitudinal. Woody stems without distinct pith, concentric woody layers and bark.

Sub-Class I.—Florideae Perigynae.

Flowers with calyx and corolla; ovary inferior, adnate to the calyx. (Corolla absent in some Hydrocharideae).

Sub-Class II.—Florideae Hypogynae.

Flowers with a calyx, corolla often absent. Stamens inserted at the base or below the ovary.

Sub-Class III.—Glumiferae.

Flowers without a conspicuous calyx, subtended by bracts. Stamens inserted at the base or below the ovary.

CLASS III.—VASCULAR ACOTYLEDONS.

No true flowers or seeds. Embryonic plant consisting of minute frond-like structure *(prothallus)* bearing male organs *(antheridia)* and female organs *(archegonia)*; the adult plant provided with leaves or fronds bearing spore-cases *(sporangia)* containing spores which originate a prothallus.

ORDERS OF PLANTS.

---o---

CLASS I.—DICOTYLEDONS.

Sub-Class I.—Choripetaleae Hypogynae.

I. Pistils separate *(apocarpous)*, each with a distinct style and stigma; seeds albuminous. *(Also Brachychiton).*

a. Carpels 2 or more. Stamens indefinite; sepals usually 5.

Herbs with radical or alternate leaves, or climbers with opposite leaves; sepals deciduous; seeds without an arillus; fruitlets not bursting, 1-seeded ...	**Ranunculaceae**
Heath-like shrubs with alternate leaves and yellow flowers; sepals 5 persistent; seeds several, with an arillus; fruitlets somewhat connate below, bursting at the top	**Dilleniaceae**

b. Carpel solitary.

Twining parasites; calyx 6-cleft; corolla 0; anthers opening by 2 or 4 valves; ovary 1-celled, 1-ovulate, pendulous; calyx enlarging over the fruit and becoming succulent	**Lauraceae**
Aquatic herb; flowers unisexual within a whorl of bracts; corolla and calyx absent; male flowers of several stamens, female of a 1-celled ovary with a pendulous ovule; fruit indehiscent ...	**Ceratophylleae**

II. Fruit of 2 or more carpels. Placentas parietal.

a. Placentas alternate with the fruit-valves.

Sepals 2, deciduous; stamens indefinite; fruit incompletely many-celled, seeds albuminous	**Papaveraceae**
Sepals and petals generally 4; stamens usually numerous; fruit 1-celled, placentas 2 or more, seeds exalbuminous	**Capparideae**
Sepals 4, deciduous; petals 4, cruciform, or rarely absent; stamens usually 4 long and 2 short; fruit 2-celled, usually bursting longitudinally by 2 valves; seeds exalbuminous	**Cruciferae**

b. Placentas opposite to the fruit-valves.

Sepals, petals, and stamens 5; petals usually irregular; filaments flat extending beyond the anthers; fruit 1-celled, placentas usually 3; seeds albuminous ... **Violaceae**

Petals and sepals 5; stamens 4 or 5; styles 2 to 5; fruit 1-celled; placentas 3; herbs beset with glandular hairs ... ... **Droseraceae**

Calyx tubular; stamens 4 to 7; petals slightly cohering; seeds albuminous; fruit 1-celled ... ... **Frankeniaceae**

III. Fruit of 2 or more carpels. Placentas axillary.

a. Fruit lobeless; calyx-lobes imbricate in the bud; no disk.

Sepals and petals 5, regular; petals rarely partially coherent; stamens 5, free; style 1; ovary 1- to 5-celled; embryo very small at the base of albumen; leaves alternate ... ... **Pittosporeae**

Sepals partly petaloid, unequal; petals unequal; stamens 8, filaments connate in two bundles; anthers 1-celled, opening by pores; fruit 2-celled; seeds 1, pendulous; embryo large, albumen scanty or 0; leaves alternate ... ... **Polygaleae**

Sepals, petals, styles, and fruit-cells 3 to 5; stamens twice as many; leaves opposite; stipules small **Elatineae**

Sepals, petals, and styles 5; petals yellow, twisted in the bud; *stamens* numerous, *connate in bundles*; herbs with opposite *dotted leaves* ... ... **Hypericineae**

b. Fruit lobed or separating into fruitlets; calyx-lobes usually imbricate. Receptacle expanded into a disk beneath the ovary, or adnate to the calyx, or rarely reduced to glands.

Petals 4 or 5, usually free; stamens twice as many; styles united; ovary raised on a fleshy disk; fruit deeply lobed or the fruitlets distinct. *Leaves with pellucid dots* ... ... **Rutaceae**

Filaments united; ovary entire, 4- to 5-celled, raised on a disk; style simple ... ... **Meliaceae**

Petals 5, twisted in the bud; stamens and staminodia 5 each, united into a ring at the base, with 5 small glands on the staminal tube opposite the petals; fruitlets separating, but leaving no central axis; ovary entire ... ... **Lineae**

Petals 4 or 5; stamens 8 or 10, free; styles united; fruit lobed, or entire in *Nitraria*; leaves stipulate; disk usually prominent ... ... **Zygophylleae**

Petals 5, twisted in the bud; stamens 10, usually connected; fruitlets separating from, or consolidated around a persistent axis; disk usually developed bearing 5 glands; leaves stipulate; herbs ... ... **Geraniaceae**

Sepals 5, imbricate or valvate in the bud; petals 5, 4, or none; stamens usually numerous; ovary usually 3-celled; disk adnate to the calyx; stipules usually 0 ... ... **Sapindaceae**

Petals 5, free or partially coherent; *stamens 5, inserted on the margin of a thin disk lining the base of the calyx*, or hypogynous; fruit separating into 3 to 5 nut-like portions from a central axis ... ... **Stackhousieae**

c. Fruit consisting of separate fruitlets; sepals imbricate; no disk.

Petals 0; fruitlets 2 or more, when many whorled, without a central axis; embryo curved around the albumen ... ... **Phytolacceae**

d. Fruit lobed or separating into distinct fruitlets, rarely entire; calyx-lobes valvate in the bud.

Petals twisted in the bud, united at the base with the staminal tube; *stamens* usually *indefinite, connate in a tube; anthers 1-celled;* fruit capsular or consisting of seceding fruitlets whorled round a common axis; seeds exalbuminous; stipules usually present ... ... **Malvaceae**

Petals 4 or 5; stamens definite, free; anthers 2-celled; ovary 3- to 5-celled; fruit not lobed (in our species); *seeds pendent*, albuminous. Stipules usually present ... ... **Tiliaceae**

Petals 5, minute, or 0; stamens definite, free or united; anthers 2-celled; ovary 3- to 5-celled; *fruit capsular or separating into distinct fruitlets* (ovary 1-celled in Brachychiton); *seeds ascending*, albuminous; stipules usually present ... ... **Sterculiaceae**

Petals 4 or 5; stamens 8 to 10, free; *anthers opening by terminal pores; fruit 2-celled, flattened, bursting at the edges;* seeds pendulous ... ... **Tremandreae**

Petals usually 0; calyx 3-partite, rarely 0; ovary 3-celled, each cell with 1 or 2 pendent ovules; *fruits separating into 3 bivalved fruitlets* from a persistent axis (1-celled, 1-seeded, opening by terminal valves in *Pseudanthus);* seeds albuminous ... ... **Euphorbiaceae**

IV. Fruit 1-celled, of 2 or more carpels.

a. Placenta central, free; seeds several.

Calyx of 2 sepals; petals 5 or more; stamens indefinite; seeds albuminous, embryo curved around the albumen; leaves alternate; (ovary half-inferior in Portulaca). ... ... **Portulaceae**

Calyx 5-cleft or of 5 sepals; petals 5, rarely 0; stamens 5 or 10; seeds albuminous, embryo curved around the albumen; leaves opposite ... **Caryophylleae**

b. Seed one in each fruit; petals usually absent (also Pseudanthus).

Calyx lobed or of distinct sepals; stamens inserted at the base of the calyx; leaves opposite ... **Illecebraceae**

Sepals 5 or 6, herbaceous or succulent; styles usually 3; embryo lateral in the albumen; *stipules united in a tube* ... ... **Polygonaceae**

Calyx usually 5-partite, herbaceous or succulent; stamens usually 5; stigmas 2 or 3; *embryo annular or coiled* ... ... **Chenopodiaceae**

Calyx 5-partite, *dry and membranous*; stigmas 1 to 3 **Amarantaceae**

Corolla tubular; calyx tubular, 5-lobed; *stamens* 5, hypogynous, *opposite the corolla-lobes*; styles 5; ovule pendulous; embryo straight in the albumen **Plumbagineae**

Calyx 5-lobed, petaloid, adherent to the fruit; *style simple*; albumen scanty, *cotyledons folded* ... **Nyctagineae**

Calyx 4- to 5-cleft; *flowers unisexual*; stamens 4 or 5; *seed pendent*, albumen scanty or wanting ... **Urticaceae**

Trees or shrubs with the ultimate branchlets cylindrical and jointed at the nodes; leaves reduced to very small scales in whorls at the nodes; flowers unisexual in separate plants; *male flowers in catkins*; stamen 1; calyx of one or two segments; female flowers without calyx; fruits seed-like, winged at the apex; the whole *fruiting mass resembling a pine-cone* (strobile) ... ... **Casuarineae**

Sub-Class II.—Choripetaleae Perigynae.

1. *Ovary superior*, quite free from the calyx. (Also Santalaceae partly, Ficoideae partly, Illecebraceae).

Ovary 1-celled, formed of 1 carpel.
Fruit a *legume*; stamens 10, rarely less, or numerous; petals present, regular or irregular, partially united or rarely wholly united; stipules usually present; leaves simple, pinnate, or absent; seeds exalbuminous ... ... **Leguminosae**

Ovary with a *pendulous ovule*; fruit not bursting; petals 0; *calyx petaloid*, tubular, *regular*, 4-lobed; *stamens* 2, alternate; flowers uni- or bi-sexual; albumen scanty or copious ... ... **Thymeleae**

Fruit a *follicle* or *berry*, 1- or 2-seeded; seeds exalbuminous, *erect;* petals 0; *calyx petaloid*, tubular, *irregular*, 4- to 5-lobed; *stamens 4 to 5 sessile* on the calyx-lobes ... ... **Proteaceae**

Ovary of 2 carpels combined at the base; styles distinct; calyx 4- to 5-cleft; petals 4, 5, or 0 ... ... **Saxifrageae**

Carpels several, distinct, each with an almost lateral style. Stamens definite; seeds albuminous ... ... **Crassulaceae**

Stamens usually indefinite; leaves *stipulate;* seeds exalbuminous ... ... **Rosaceae**

 II. Ovary inferior, syncarpous; *stamens inserted on the calyx.* (Also Portulaca).

  a. Placentas parietal.

Calyx-tube adnate to the ovary, or if free from it with a distinct tube bearing stamens; stamens few or indefinite; ovary, cells and styles usually 3 to 5; embryo around the albumen ... ... **Ficoideae**

  b. Placentas axillary.

Ovary adnate to the calyx-tube at the base only; 2- to 4-celled; style simple; seeds numerous, exalbuminous **Lythraceae**

Fruit adnate to the calyx-tube high up beyond its base. Stamens 8; petals 4; seeds numerous, exalbuminous **Onagreae**

Stamens usually indefinite; *leaves* exstipulate, *transparently dotted;* seeds exalbuminous ... ... **Myrtaceae**

 III. Ovary inferior, or half inferior; *stamens inserted* on the *margin of a disk* lining the calyx-tube.

Ovary 3-celled; *ovules solitary, erect;* petals minute concave or 0; stamens 5, alternate with the calyx-lobes **Rhamneae**

Ovary 1-celled; ovules 3, pendulous; petals free or slightly connate; calyx-lobed inclosed within a *calyciform involucre;* stamens 3, alternate with the calyx-lobes; staminodia present ... ... **Olacineae**

Ovary 1-celled; ovules 1 to 5, pendulous; calyx 5-lobed; *stamens 5, opposite* to the calyx-lobes; fruit drupaceous with one erect seed ... ... **Santalaceae**

 IV. Ovary inferior; 2- to 4-celled with *separate styles and stigmas;* seeds solitary, pendulous, albuminous.

Stamens usually twice as many as petals or calyx-lobes; fruits of 2 to 4 connate fruitlets ... ... **Halorageae**

Stamens 5, opposite to the usually small incurved petals; *stamens and petals inserted on a terminal (epigynous) disk;* fruit of two connate fruitlets, usually separating and often leaving a persistent filiform axis *(carpophore);* very rarely reduced to 1 carpel ... **Umbelliferae**

Sub-Class III.—Synpetaleae Perigynae.

(Also Primulaceae partly).

Fruit fleshy with 3 *parietal placentas*; flowers unisexual, males usually clustered or racemose, female solitary; corolla-base confluent with the calyx; stamens 5 in 3 parcels; anthers large on short filaments twisted or straight; *climbing or trailing herbs* by the aid of lateral unbranched tendrils ... ... **Cucurbitaceae**

Parasitic shrubs; ovary 1-celled, 1-ovulate; fruit a drupe; petals 5 or 6, free or partially united; *stamens* 5 or 6, adnate and *opposite to the corolla-lobes*; calyx without lobes ... ... **Loranthaceae**

Leaves opposite or whorled, *simple*; stamens 4 or 5 alternate; *fruit 2-celled*; stipules present ... **Rubiaceae**

Leaves opposite, pinnate; stamens 3 to 10, alternate; ovary 3- to 5-celled with one pendulous ovule in each; fruit a berry ... ... **Caprifoliaceae**

Stamens 5 united by their anthers; stigma bifid; *flowers in heads with an involucre of bracts* (termed phyllaries); *fruit 1-celled, 1-seeded*; calyx-limb reduced to hairs or scales (termed pappus), or wanting ... ... **Compositae**

Stamens 2, connate with the style into a "column;" corolla 5-lobed, the 5th lobe usually very small; fruit 2-celled ... ... **Candolleaceae**

Stamens 5, free or synantherous, calyx-lobed. Stigma lobed, fruit 2- or more-celled ... ... **Campanulaceae**

Stigma concave with a more or less cup-shaped ciliate membrane (indusium); fruit 1- or 2-celled **Goodeniaceae**

Sub-Class IV.—Synpetaleae Hypogynae.

1. Corolla regular, stamens equal in number with the petals (except Jasminum and Solanaceae partly).

a. Anthers 2-celled.

Stamens opposite to the corolla-lobes (also Plumbagineae).
Fruit capsular, many seeded, placenta free ... **Primulaceae**
Stamens alternate with the corolla-lobes.
Fruit lobed or separating into distinct fruitlets.
Fruit 2- or rarely 1-celled; seeds few, erect; placentas basilary; embryo twisted or folded ... ... **Convolvulaceae**

Fruit 4-lobed, or of 2 or 4 fruitlets, each
with 1 pendent seed; leaves alternate,
usually hispid ... ... **Boragineae**

Fruitlets 2; *anthers connate*, with dorsal
appendages *(corona); pollen consolidated*,
affixed to 5 distinct processes of the stigma
placed between the anthers; seeds numer-
ous, hair-tufted; stems with milky juice **Asclepiadeae**

Fruitlets 2, seldom with a 2-celled fruit;
anthers connate; pollen powdery ... ... **Apocyneae**

Fruit lobeless of 2 united carpels, embryo straight
(also Bruonia).
   Placentas 2, parietal; leaves opposite ... **Gentianeae**

Placentas 2, basal or axillary.
   Stamens 2; seeds few; leaves opposite **Jasmineae**

   Stamens 4; seeds few; corolla-lobes 4,
imbricate in the bud; capsule burst-
ing transversely; leaves radical ... **Plantagineae**

   Stamens 4 or 5; seeds many; leaves
opposite, sometimes stipulate ... **Loganiaceae**

Fruit lobeless of 2 united carpels, embryo curved;
corolla-lobes 5, valvate or folded in the bud;
fruit 2-celled; placentas axillary ... ... **Solanaceae**

      b. Anthers 1-celled (also Anthotriche).

Stamens usually 5; fruit 2- to 5-celled, few or many
seeded; *leaves stiff*, scattered; placentas axillary **Epacrideae**

   II. Corolla irregular usually bilobed; stamens fewer than the
corolla-lobes; alternate, 2 or 4, if the latter usually in
pairs 2 long and 2 short.

Fruit 4-lobed or separating into 4 indehiscent nuts.
   Leaves opposite; herbaceous stems quadrangular; usually pubescent; corolla bilobed ... **Labiatae**

Fruit lobeless, syncarpous.
   Ovary 1-celled, many ovules, corolla bilobed.
     Placentas free central, stamens 2, stigmas 2,
leaves radical ... ... **Lentibularineae**

     Placentas parietal; stamens 4; herbaceous
parasites, leaves reduced to scales ... **Orobancheae**

   Ovary 2-celled, many ovules, placentas axilliary
(also Goodeniaceae partly).
     Seeds small, minute, albuminous; leaves op-
posite or alternate; stamens 2 or 4; cor-
olla usually 2-lobed ... ... **Scrophularineae**

Seeds large, *winged*, exalbuminous; stamens in pairs; leaves opposite; corolla bilobed ... **Bignoniaceae**

Ovary 2- to 4-celled, ovules few.
Ovary 2-celled, 2 or more ovules in each; seeds exalbuminous; leaves opposite ... **Acanthaceae**

Ovary 2-celled, but divided in 4 cells by spurious dissepiments, one erect ovule in each ... **Pedalineae**

Fruit 2- to 4-celled; seeds albuminous.
Fruit 4-celled, drupaceous or separating into fruitlets, rarely dehiscent; seeds erect, 1 in each cell; leaves usually opposite; stigma bilobed ... **Verbenaceae**

Fruit indehiscent, seeds pendulous, 1 to 4 in each cell; leaves usually alternate, often dotted; stigma simple ... **Myoporineae**

Sub-Class V.—Gymnosperms.

Branching trees, leaves scale-like whorled or opposite; stamens in catkins, sessile on dilated scale-like bracts; ovules in the axils of a cone ... **Coniferae**

Trunks simple with a palm-like crown of large pinnate leaves; anthers on the under side of the scales of a large cone; ovules in the axils of a cone ... **Cycadeae**

CLASS II.—MONOCOTYLEDONS.

Sub-Class I.—Florideae Perigynae.

Flowers regular, mostly unisexual; sepals herbaceous; petals coloured, often very tender and fugacious, or absent; fruit not bursting, placentas parietal; aquatic herbs... **Hydrocharideae**

Flowers irregular, the lower petal (labellum) usually unlike the two others; sepals 3, usually petaloid; anther one, on a central *column* bearing the stigmas which are confluent in a mucous disk; pollen in masses; fruit dehiscent, placentas 3 parietal; terrestrial herbs with tuberous rootstock; rarely epiphytic ... **Orchideae**

Flowers regular of 6 petaloid parts.
Stamens 3, opposite the calyx-lobes; anthers extrorse ... **Irideae**

Stamens 6, anthers bursting inwards ... **Amaryllideae**

Sub-Class II.—Florideae Hypogynae.

I. Petals coloured.

Fruit entire, 3-celled (rarely 1-celled); placentas axillary; sepals generally petaloid; stamens 6, anthers opening by longitudinal slits, rarely by pores; stigma 3-lobed or entire; embryo within the albumen ... **Liliaceae**

Fruit entire, 1-celled, or imperfectly 3-celled; placentas 3, parietal; petals connate; fertile stamens 3 opposite the petals, sterile ones 3, bearded; stigmas 3 ... **Xyrideae**

Fruit entire, 3-celled; sepals 3, herbaceous; petals 3, delicate, spreading; style and stigma simple ... **Commelineae**

Fruit consisting of distinct fruitlets; petals white, very deciduous; seeds exalbuminous ... ... **Alismaceae**

II. Petals sepal-like.

Herbs with grass- or rush-like leaves; no spathe; style with 3 linear branches; fruit capsular ... ... **Juncaceae**

Arborescent, trunks simple with a crown of large pinnate leaves; young inflorescence enclosed in a spathe ... **Palmae**

III. Sepals and petals reduced to scales or none.

Flowers in dense elongated spikes, the upper ones male, the lower female; sepals and petals pappus-like; tall semi-aquatic herbs with very long leaves ... **Typhaceae**

Aquatic, rarely terrestrial herbs; flowers solitary or clustered; fruit consisting of distinct or connate fruitlets ... ... **Fluviales**

Floating plants consisting of minute green scale-like fronds without stem or leaves; flowers reduced to an anther and an ovary ... ... **Lemnaceae**

IV. Sepals and petals bract-like, rarely absent; embryo outside the albumen; seed pendulous.

Leaves often rudimentary, rigid or reduced to sheathing scales; male and female flowers mostly in separate plants; sepals and petals 3 or less; fruit 1- to 3-celled; rush- or sedge-like plants ... ... **Restiaceae**

Sub-Class III.—Glumiferae.

Clasping leaf-stalks tubular, with connate margins; stems solid, without nodes, often angular; floral bracts solitary; floral segments none or of small hypogynous scales or bristles; style 1, or 2- or 3-cleft; stamens 3 to 12 ... ... **Cyperaceae**

Clasping leaf-stalks with free margins; stems usually hollow, jointed, round; flowers in more or less scarious spikes called *spikelets*; 2 or 3 scale-like bracts called *glumes* subtending the spikelet; true floral-segments usually absent, rarely of 3 pellucid scales called *lodicules*; each flower usually enclosed in a 2-nerved scale called a *palea* (regarded as 2 connate bracteoles) and an outer scale or *flowering glume*; styles 2, usually feathery; stamens 1 to 4, usually 3 **Gramineae**

CLASS III.—VASCULAR ACOTYLEDONS.

Spore-cases in spikes, supported by bracts, in the axils of leaves or at the summit of the branches ... **Lycopodiaceae**

No true leaves; foliaceous organs or *fronds* circinate in vernation.
    Barren fronds linear or with leaf-like laminae; fertile fronds, often emanating at or near the roots, forming a closed involucre including the spore-cases, containing spores of two kinds ... ... **Rhizospermae**

    Fertile fronds bearing the spore-cases on their under side or margins; spore-cases stalked or sessile, with or without an encircling elastic ring, opening by regular slits, or by rupture ... ... **Filices**

GENERA AND SPECIES OF PLANTS.

CLASS I.—DICOTYLEDONS.

Sub-Class I.—Choripetaleae Hypogynae.

ORDER RANUNCULACEAE.

Petals, 0; fruits with feathery styles; sepals white, valvate in bud; climbing shrubs with opposite compound leaves	**Clematis**
Petals, 5 to 12; sepals imbricate in bud; herbs with radical or alternate leaves.	
Carpels collected into a globular mass, ovule ascending; petals with a nectar gland	**Ranunculus**
Carpels imbricate on a long receptacle; ovule pendulous	**Myosurus**

Clematis.

Anthers with long appendages; leaves somewhat rigid ...	*aristata*
Anthers short, without appendages; leaves rather flaccid	*microphylla*

Ranunculus.

Carpels wrinkled; petals white, no nectar scale. Floating in water; leaves submerged, finely divided	*aquatilis*
Carpels smooth; petals yellow, with a nectar scale.	
Stem tufted, hairy; petals, 5; sepals appressed; style recurved	*lappaceus*
Stem creeping; petals 5-12; style straight; marsh plant	*rivularis*
Carpels rough; flowers small, lateral, sessile. Dwarf annual	*parviflorus*

Myosurus.

Leaves radical, linear; stamens 5 to 20, scapes one-flowered. Annual	*minimus*

ORDER DILLENIACEAE.
Hibbertia.

I. Stamens on one side of the carpels, all fertile.

Flowers nearly sessile.
    Sepals and floral leaves 2 lines long; petals narrow ... *hirsuta*
    Sepals 3 to 5 lines long.
        Leaves soft, hairy; floral leaves crowded, long; petals broadly cordate; outer sepals somewhat silky ... *sericea*
        Leaves scabrous, almost linear; flowers scattered ... *stricta*
Flowers distinctly stalked.
    Leaves obovate, cuneate at the base; stamens 10-12 ... *Billiardieri*
    Leaves linear, glabrous, sharp-pointed; stamens 8 or less *acicularis*

II. Stamens under 20, all round the carpels, all fertile.

Bracts scarious, very broad; leaves narrow-linear, glabrous *virgata*
Bracts small, sepal-like; leaves linear, clustered, hairy ... *fasciculata*

III. Stamens 200 to 300, all round the carpels, with 20 or more sterile ones outside.

Leaves oblong-lanceolate, almost clasping; glabrous... ... *glaberrima*

ORDER LAURACEAE.
Cassytha.

Wiry, twining stems; leaves scale-like; flowers small in spikes.
    Flowers in globular clusters, few, glabrous; fruit ovoid and glabrous; stems threadlike ... ... *glabella*
    Flowers pubescent, in globular clusters when young, afterwards elongate; fruit globular downy; stems moderate, smooth or warty ... ... *pubescens*
    Flowers in very short spikes, large, glabrous, drying black; fruit globular thick; stem thick ... ... *melantha*

ORDER CERATOPHYLLEAE.
Ceratophyllum.

Leaves whorled, dichotomously divided into linear segments *demersum*

ORDER PAPAVERACEAE.
Papaver.

Erect annual with milky juice, beset with bristly hairs; leaves lobed; petals large, red; capsule glabrous, ovoid-oblong, opening by pores beneath the disk-like summit... *aculeatum*

ORDER CAPPARIDEAE.

Herbs with a capsular fruit, stamens 8 to 16 ... ... **Cleome**
Shrubs or trees; fruit indehiscent, succulent, on a long stalk; stamens indefinite; stipules spinescent, leaves simple ... **Capparis**

Cleome.

Erect, branching about 1 foot, beset with viscid hairs; leaves of 3 or more obovate leaflets; flowers yellow in terminal racemes ... ... *viscosa*

Capparis.

Sepals 4, imbricate in two series.
    Stamens 12 or less; branchlets and inflorescence tomentose *lasiantha*
    Stamens indefinite; glabrous, prostrate shrub ... ... *spinosa*
Two outer sepals connate, bursting irregularly. A small tree resembling the orange; fruit globose, with a hard rind ... *Mitchelli*

ORDER CRUCIFERAE.

Pod longer than broad, septum in its broadest diameter, separating into two valves from below upwards.

    a. Valves nerveless (also Stenopetalum partly).

Pod cylindrical, sepals spreading; seeds in 2 rows **Nasturtium**
Pod compressed, sepals erect; seeds usually in 1 row **Cardamine**

    b. Valves 1-nerved; seeds numerous.

Pod quadrangular, sepals erect; seeds in 1 row ... **Barbarea**
Pod cylindrical, sepals erect, hairy; seeds in 1 or 2 irregular rows ... ... **Erysimum**

    c. Valves 3-nerved; seeds numerous.

Pod narrow-cylindrical, sepals erect or spreading; seeds in 1 or 2 irregular rows in each cell ... **Sisymbrium**
Pod longer than broad, separating transversely into two portions, each 1-seeded ... ... **Cakile**
Pod shorter than broad, septum in its broadest diameter, 2-valved.
    Sepals coherent; *petals long subulate-pointed;* leaves linear; seeds few or several ... ... **Stenopetalum**
    Fruiting-stalks recurved; flowers minute ... ... **Geococcus**
    Filaments toothed; seeds 2 to 4 in a cell ... ... **Alyssum**
    Dissepiment absent; seeds numerous; small annuals with linear entire leaves and small flowers ... **Menkea**

Pod shorter than broad, septum in its narrow diameter.
Seeds 2 or more in a cell; pod ovoid or compressed — **Capsella**
Seeds 1 in each cell; pod compressed, obcordate … — **Lepidium**

Nasturtium.

Marsh plant. Leaves pinnatifid; flowers yellow in short racemes … … … … … … *terrestre*

Cardamine.

Pod nearly cylindrical, seed in rows; style long … *eustylis*
Pod compressed, seeds in 1 row in each cell, style short.
  Petals narrow, erect, scarcely longer than the calyx; stems erect, almost leafless; stamens 4. Marsh plant … … … … … … *laciniata*
  Petals larger, obovate spreading; seeds as broad as the septum; stamens 6. Slender branched annual … … … … … … … *flexuosa*

Barbarea.

Erect stout herb, radical leaves pinnatifid with a large terminal lobe; flowers yellow … … … *vulgaris*

Erysimum.

Petals scarcely exceeding the calyx. Hoary annuals.
  Pedicels spreading or curved, as long as the pod; flowers yellow. Dwarf annual, leaves lanceolate *curvipes*
  Pedicels erect, shorter than the pod; flowers white or pink … … … … … … *brevipes*
Petals twice as long as the calyx; hoary with stellate pubescence; flowers white or pink.
  Pod lanceolate, hairy; calyx 1 line long; seeds few *lasiocarpum*
  Pod linear, slightly pubescent; calyx 2 to 5 lines; seeds many; leaves pinnatifid or incised … *Blennodia*

Sisymbrium.

Shrubby perennials; leaves or their lobes linear-filiform, glabrous.
  Leaves entire; flowers white. A small shrub … *filifolium*
  Leaves divided into 3 segments; flowers pale yellow; stems herbaceous from a woody base … *trisectum*
Annuals; leaves lobed, glabrous; flowers small; yellow.
  Erect; fruiting pedicels erect; leaves pinnately divided into a few linear segments; pod narrow *nasturtioides*
  Prostrate, dwarf; fruiting pedicels spreading; leaves oblong, coarsely toothed or shortly pinnatifid; pod broad … … … … … *procumbens*

Annual; invested with simple appressed hairs; leaves pinnatifid or incised; pod linear; flowers yellow... *Richardsii*

Annual; leaves pinnatifid, stellately pubescent; pod narrowed toward the base; flowers white... *cardaminoides*

Cakile.

A coarse glabrous herb, inhabiting sandy sea-shores ... *maritima*

Stenopetalum.

Pods erect, elongate, 2 to 5 times as long as broad.
    Hoary tomentose; pedicels as long as the pod; petals thrice as long as the calyx ... *velutinum*
    Glabrous; pedicels shorter than the pod; petals yellow, twice as long as calyx; leaves few, narrow-linear ... *lineare*

Pods spreading or pendulous, globular or ovoid.
    Hirsute; pedicels slender, 2 or 3 times longer than calyx; petals 4 or 5, more than twice as long as calyx; pod oval-oblong ... *nutans*
    Glabrous; pedicels 2 to 3 lines long; petals under 2 lines, scarcely longer than calyx; pod globular *sphaerocarpum*
    Glabrous; pedicels shorter than sepals; *petals* yellow, with *long trisect points;* pod nearly globular ... *croceum*

Geococcus.

Dwarf, stemless, tufted annual; with pinnately divided spreading leaves, 1½ to 3 inches long ... *pusillus*

Alyssum.

Dwarf, wiry, erect, hoary annual; leaves linear to oblong-spathulate; flowers white, very small ... *minimum*

Menkea.

Pod ovate, 2 lines long; petals white ... *australis*
Pod globular, 1½ to 2 lines long; petals yellow ... *sphaerocarpa*

Capsella.

Pod laterally compressed, cuneate, *emarginate* atop.
    Dwarf, erect, much branched, hairy annual; leaves small, obovate; flowers white ... *pilosula*

Pod elliptical or ovoid, not compressed.
    Glabrous, dwarf, slender, decumbent annual; flowers white; seeds 10 to 12 in each cell ... *elliptica*
    Glabrous, prostrate; flowers yellow; seeds 4 ... *humistrata*

Hairy, erect, rigid, branching; flowers white;
leaves lanceolate, entire, stalked... ... ... *cochlearina*

Stellately pubescent; stem more branched ... *Drummondi*

Lepidium.

Leaves all entire; pod usually conspicuously winged.
  Shrubby, much branched, glabrous.
    Leaves broadly ovate or orbicular ... ... *strongylophyllum*

    Leaves narrow-linear; petals linear, white;
    stamens of equal length... ... ... *leptopetalum*

  Herbaceous; leaves linear; petals oblong to ovate, or 0.
    Pod winged to the base, the lobes longer than
    the style; petals white ... ... *rotundum*

    Pod scarcely winged; style small, slender, but
    longer than the notch; petals pale-lilac ... *phlebopetalum*

    Pod winged to the base, the lobes almost united
    to the style; petals 0; *stamens* 4 ... ... *monoplocoides*

Leaves toothed or lobed; pod-wings small or 0; herbs, more or less glabrous.
  Stems beset with papillae; upper leaves auricled;
  pod shortly winged; petals 0; stamens 4 ... *papillosum*

  Stems glabrous or slightly hairy; pod scarcely
  winged, minutely lobed at the top; stamens 2;
  petals 0. Sometimes with corymbose racemes
  and spinescent branchlets ... ... *ruderale*

  Stems glabrous; pod with narrow wings, distinctly
  lobed at the top; petals 4, minute; stamens 6 ... *foliosum*

ORDER VIOLACEAE.

Flowers irregular; fruit capsular.
  Sepals spurred or protuberant at the base; lower
  petal spurred or saccate at the base ... ... **Viola**

  Sepals not produced at the base; lower petal
  larger, gibbous or saccate at the base ... ... **Hybanthus**

Flowers regular; fruit a berry; *anthers united* ... ... **Hymenanthera**

Viola.

Stemless with rooting offshoots; leaves reniform or
orbicular; flowers violet, small; stipules free ... *hederacea*

Stemless, no stolons; leaves lanceolate to oblong;
flowers large; stipules linear, adnate to the petiole *betonicaefolia*

GENERA AND SPECIES.

Hybanthus.

Low undershrubs, peduncles not longer than the leaves.
  Peduncles 2- to 4-flowered, flowers blue and white; lowest petal small, distinctly clawed; leaves alternate, linear ... ... *floribundus*
  Peduncles 1-flowered; lowest petal more than twice as long as the calyx ... ... *enneaspermus*
Glabrous, slender, dwarf herb; peduncles slender, much longer than the linear leaves... ... ... *Tatei*

Hymenanthera.

An intricately branched thorny shrub; leaves stiff, oblong-elliptical to linear, distantly toothed; flowers small, green, axillary ... ... *Banksii*

ORDER DROSERACEAE.

Drosera.

Root fibrous or bulbous; leaves radical or along the flower-stalks; styles simple or divided into filiform branches.

              I. Leaves radical; scapes leafless.

Leaves several inches long, divided into 2 long linear lobes, on long petioles; flowers large, pink or white, in 2 or 3 racemose branches; styles 3, divided into numerous forked branches; tall plant ... ... *binata*
Leaves more or less ovate or orbicular.
  Stipules absent, root fibrous; flowers crimson, small in glandular-hairy racemes; styles 3, each 2-branched; dwarf plant ... ... *glanduligera*
  Stipules absent; root bulbous; petals white; flowers large, solitary, on rather short scapes; styles 3, divided into numerous branches ... ... *Whittakeri*
  Stipules scarious, lobed.
    Scapes 1-flowered; sepals, petals (white), stamens, and styles 4; root fibrous; a minute annual ... *pygmaea*
    Scapes racemose.
      Leaves ovate or spathulate; flowers small; petals red to white; root fibrous; styles 3 or 4 divided to the base into 2 branches ... *spathulata*
      Leaves broader; styles 5, not branched ... *Burmanni*

            II. Leaves on the stem, with or without basal leaves.

Leaves linear, several inches long; flowers in lateral racemes; styles 3, bifid; root fibrous; tall plant ... *Indica*

Leaves on the stems peltate; lower leaves reduced to acute scales; flowers large pink, few in a short raceme; styles divided into numerous forked branches; stems slender trailing; root bulbous ... ... *Menziesi*

Leaves on the stem orbicular-reniform; lower leaves rosulate-spreading; styles 3, divided into numerous forked branches; root bulbous; stems erect, simple or slightly branched, about 1ft.; flowers white.
    Sepals entire, glabrous; seeds narrow linear, *the loose testa produced beyond the nucleus* ... ... *auriculata*
    Sepals toothed, hairy, closely appressed; seeds ovoid ... ... *peltata*

ORDER FRANKENIACEAE.

Frankenia.

Procumbent undershrub; leaves small, linear to ovate-lanceolate; flowers pink, scattered or forming a leafy cyme   *laevis*

ORDER PITTOSPOREAE.

Anthers ovate or oblong, bursting lengthwise.
  Capsule bursting, of thick consistence; seeds enveloped in a sticky fluid ... ... **Pittosporum**
  Capsule bursting, of thin consistence.
    Petals spreading; seeds vertical, flat ... ... **Bursaria**
    Petals partially cohering; seeds horizontal, thick ... ... **Marianthus**
  Fruit an ovoid or oblong berry ... ... **Billardiera**
Anthers linear, turned to one side, opening in terminal pores ... ... **Cheiranthera**

Pittosporum.

A small tree with drooping branches; pedicels axillary, flowers yellow; leaves linear-oblong, flat ... ... *phillyraeoides*

Bursaria.

A prickly shrub or small tree; flowers white in terminal panicles; capsule flat, broadly orbicular; leaves small, oblong ... ... *spinosa*

Marianthus.

A slender twiner, leaves stalked, oblong, about 1 inch; flowers orange, solitary, axillary, about 1 inch ... *bignoniaceus*

Billardiera.

Stems twining; pedicels solitary, flowers yellow; leaves ovate-linear, wavy on the margin; sepals lanceolate   *scandens*

A small shrub, with the branches sometimes twining; flowers blue in corymbs or sessile clusters; leaves lanceolate, flat; sepals lanceolate or ovate ... ... *cymosa*

Cheiranthera.

An erect glabrous undershrub about 1ft.; flowers blue, large, in corymbs; leaves linear, usually flat ... *linearis*

A slender twiner; flowers blue, solitary, stalked; leaves linear with involute margins ... ... *volubilis*

ORDER POLYGALEAE.

Inner sepals large and petal-like; anthers 8.
    Lateral petals united with the crested lower petal; capsule sessile; seeds hairy or glabrous ... ... **Polygala**
    Lateral petals united to the staminal column, but distinct from the plain lower petal; capsule usually stalked; seeds hairy or hair-tufted *(coma)* ... **Comesperma**

Polygala.

An annual with stalked orbicular leaves; flowers blue ... *Chinensis*

Comesperma.

Capsule sessile, cuneate-obovate; seeds slightly hairy.
    Small shrub with erect, rigid, leafless branches ... *scoparium*
Capsule narrowed at the base; seeds with a long coma.
    Outer sepals free, shorter than the inner.
        Stems twining, almost leafless; flowers blue or white in axillary or terminal racemes... ... *volubile*
        Erect shrub; leaves glaucous, oblong, mucronate *sylvestre*
        Erect viscid shrub; leaves lanceolate-ovate; inner petals yellow ... ... *viscidulum*
    Outer sepals free, as long as the inner.
        Glabrous perennial with erect leafy stems about 1ft.; leaves linear to elliptical or oblong; flowers small, blue, in slender racemes ... *calymega*
    Two outer sepals united; flowers pink; otherwise very similar to *C. calymega* ... ... *polygaloides*

ORDER ELATINEAE.

Sepals membranous, blunt; outer portion of fruit membranous; floral parts in 3's ... ... **Elatine**

Sepals herbaceous, pointed; outer portion of fruit rather hard; floral parts in 4's or 5's ... ... **Bergia**

Elatine.

A small tender glabrous annual prostrate or creeping over mud; leaves ovate to broadly oblong; flowers solitary, axillary; seeds curved, wrinkled ... ... *Americana*

Bergia.

Stamens as many as the petals or sepals; flowers clustered, axillary; small pubescent or hairy annual ... ... *ammannioides*

Stamens twice as many as the petals or sepals; flowers solitary, stalked; a glabrous or slightly hairy perennial with woody prostrate branches; leaves ovate ... *perennis*

ORDER HYPERICINEAE.

Hypericum.

Small erect glabrous herb; leaves from oval to lanceolate-elliptical; sepals lanceolate; fruit oval-ellipsoid, 1-celled, 3-valved ... ... *Japonicum*

ORDER RUTACEAE.

I. Leaves opposite; petals 4, united or free.

Petals 4, united into a cylindrical corolla; calyx cup-shaped and undivided, or cleft; undershrubs with simple stalked leaves and large showy pendulous flowers ... ... **Correa**

Petals 4, free, spreading; calyx 4-cleft.
   Stamens 4, inserted on the outside of distinct gland-like bodies, alternating with the petals ... ... **Zieria**
   Stamens 8; disk without glands; leaves simple or compound; undershrubs or almost herbaceous ... **Boronia**

II. Leaves alternate, simple; petals 5, free.

Stamens usually 10; fruitlets usually 5, pointed; shrubs. **Eriostemon**
Stamens 5; fruitlets 5, blunt; small trees or shrubs ... **Geijera**

Correa.

Calyx cleft, the lanceolate teeth as long as the tube.
   Petals separating after the flower is expanded, green or purple; filaments dilated at the base, anthers yellow ... ... *aemula*

Calyx truncate, with four minute teeth.
   Petals separating after the flower is expanded, white or pink; filaments filiform, anthers red ... ... *alba*

Petals cohering till the flower falls, red, white, or yellowish-green; four of the filaments dilated below the middle, anthers yellow; branchlets, leaves, and inflorescence more or less clothed with stellate hairs; leaves broadly ovate or cordate to narrow-oblong ... ... *speciosa*

Calyx truncate with 4 short broad, and 4 longer filiform teeth. Petals cohering; the filaments all slightly dilated below the middle. A decumbent stellately tomentose shrub ... ... *decumbens*

Zieria.

Low erect shrub; leaves softly tomentose, oblong; flowers 1 to 3 on short stalks, white, small, axillary *veronicea*

Boronia.

I. Petals valvate in the bud.

Small, erect, much branched shrub; *leaflets* 3, small, oblong, *flat;* flowers pink, 1 to 3 together on slender stalks, terminal or axillary; *anthers tipped with recurved points* ... ... *Edwardsi*

II. Petals imbricate in the bud.

Leaves or leaflets filiform or semicylindrical.
  Leaves simple; flowers blue, axillary; filaments ciliate; anthers with short, broad, obtuse, recurved appendages ... ... *coerulescens*

  Leaves of 3 to 5 leaflets; flowers pink or red, terminal. Anthers without appendages; filaments slightly ciliate; leaves very narrow, simple and sessile, or consisting of 3 linear leaflets; petals pink; seeds shining ... ... *filifolia*

  Anthers with small appendages, filaments glabrous; leaflets 3 to 5, small, clavate-cylindrical, clustered on very short stalks; petals crimson; seeds smooth but not shining ... ... *clavellifolia*

Leaves or leaflets flat.
  Leaves simple; sepals nearly as long as the pink or whitish corolla; filaments slightly hairy, anthers without appendages; seeds shining. Dwarf, almost herbaceous ... ... *parviflora*

  Leaves mostly of *3* linear-oblong or obovate *leaflets*, sometimes simple; sepals much shorter than the pink petals; filaments hairy, anthers with appendages; seeds rough not shining ... ... *polygalifolia*

Leaves pinnate of several pairs of linear- to oblong-lanceolate leaflets; petals large pink; filaments hairy, anthers inconspicuously appendaged; seeds smooth ... *pinnata*

Eriostemon.

I. Petals imbricate in the bud. Inflorescence without scurfy scales; carpels 5.

Flowers axillary, usually solitary.
  Leaves obovate or spathulate, thick, of a greyish hue; filaments flattened, ciliate ... *obovalis*
  Leaves narrow-linear; filaments filiform, hairy ... *linearis*

Flowers terminal, usually 2 or 3 together. Leaves small, flat, or linear-terete, usually tuberculate; filaments flattened, ciliate ... *difformis*

II. Petals slightly imbricate with inflexed valvate tips. Inflorescence umbellate. Beset with scurfy scales.

Leaves oblong or linear, rounded or obtuse $\frac{1}{2}$ to $1\frac{1}{2}$in. ... *lepidotus*

Leaves linear-cuneate, margins revolute or recurved; bi-lobed at the summit, under 1in. ... *sediflorus*

III. Petals valvate glabrous, no scurfy scales; carpels 5.

Flowers axillary, solitary, short-stalked, white, glabrous; stamens included; leaves linear or linear-lanceolate, rigid, pungent-pointed ... *pungens*

Flowers in terminal clusters; stamens exsert.
  Leaves oblong or lanceolate, truncate or 2-lobed at the end; flowers small in nearly sessile umbels ... *Hillebrandi*
  Leaves very small obovate thick very obtuse and convex; flowers small in clusters of 3 to 5... *brachyphyllum*

IV. Petals slightly imbricate. Carpels 2.

Branchlets and underside of leaves beset with scurfy scales; leaves spreading, linear, obtuse with revolute margins; flowers small in dense sessile heads, amongst the uppermost leaves ... *capitatus*

Geijera.

Moderate-sized tree with drooping foliage, leaves lanceolate 3 to 6 inches; panicle loose, many-flowered ... *salicifolia*

Tall shrub, leaves linear, thick, obtuse; panicle short and few-flowered ... *parviflora*

ORDER MELIACEAE.
Owenia.

Stamens 10, staminal tube toothed between the anthers; ovules solitary in each cell; fruit drupaceous.
A small tree, leaves pinnate; leaflets numerous lanceolate, acute, 1-nerved ... ... *acidula*

ORDER LINEAE.
Linum.

Erect glabrous herb; petals blue without appendages; styles 5 united to the middle; leaves narrow ... ... *marginale*

ORDER ZYGOPHYLLEAE.

Leaves pinnate, flowers solitary, fruit of hard indehiscent coherent nuts. Prostrate hairy herbs ... **Tribulus**

Leaves of two succulent leaflets or lobes; flowers solitary; petals 4 or 5, flat, thin, usually yellow; fruit 3- to 5-angled or-lobed, bursting longitudinally or indehiscent ... ... **Zygophyllum**

Leaves simple succulent; flowers in cymes; petals concave; fruit a drupe ... ... **Nitraria**

Tribulus.

Each fruitlet rounded at the back, prickly.
   Leaflets small obliquely oblong, in 4 to 8 pairs; flowers usually small; stamens 10; fruitlets 5, with 2 marginal conical prickles ... ... *terrestris*
      Flowers usually larger; fruitlets covered with numerous nearly equal prickles ... ... *hystrix*
Each fruitlet with prominent almost winged angles and 2 dorsal prickles ... ... *macrocarpus*
Fruit with 5 wingless rays near the summit, and 5 basal tubercles ... ... *astrocarpus*

Zygophyllum.

i. Capsule truncate at the top.

Sepals and petals 5; filaments broadly winged to the middle; angles of the capsules produced into blunt appendages; leaflets ovate, oblique ... ... *apiculatum*
Sepals and petals 4; filaments not winged.
   Angles of the capsule with membranous wings.
      Flowers large yellow, stamens 8; seeds 4 to 6 in a cell. A low lax shrub ... ... *fruticulosum*

Angles of the capsule acute, not winged.
  Flowers minute white, stamens 4, seeds 2 in a
    cell. Dwarf annual ... ... *ammophilum*
  Flowers large yellow; stamens 8. Diffuse
    trailing undershrub, leaflets elliptical to
    linear ... ... *Billardieri*

II. Capsule oval.

Sepals and petals 4; stamens 8.
  Angles of the capsule terminating in small erect
    leafy appendages, filaments with short narrow
    entire wings ... ... *prismatothecum*
  Angles of the capsule 3, winged; filaments not
    winged; flowers minute ... ... *Howitti*
  Angles of the capsule thick and narrow, filaments
    with toothed wings.
      Leaflets cuneate-obovate entire; flowers large.
        Erect robust herb ... ... *glaucescens*
      Leaflets broadly cuneate, notched at the end;
        flowers rather small. Small annual ... *crenatum*
Sepals and petals 5, stamens 10, capsule bluntly 5-
  angled; filaments with short narrow wings; leaflets
  oblong-cuneate, notched at the end. Diffuse
  annual, fruit often assuming a violet hue ... ... *iodocarpum*

Nitraria.

A rigid spreading shrub, branchlets often spinescent.
  Fruit yellow ripening to dark-purple. Saline
  tracts ... ... *Schoeberi*

ORDER GERANIACEAE.

Capsule separating into 5-beaked, 1-seeded, fruitlets.
  Leaves lobed; flowers umbellate.
    Petals unequal; a nectar-tube adnate to the
      pedicel ... ... **Pelargonium**
    Petals equal; no nectar-tube.
      Beaks of fruitlets glabrous inside; stamens 10 **Geranium**
      Beaks of fruitlets bearded inside; stamens 5 **Erodium**
Capsular valves adherent to the axis; leaflets 3 ... ... **Oxalis**

Pelargonium.

Leafy stems elongate; leaves reniform-cordate, crenate,
  or shortly lobed; peduncles usually longer than the
  leaves; petals whitish or pink not twice the length
  of calyx ... ... *australe*
Stems short, leaves chiefly radical, rhomboid-ovate;
  petals rose-colored, much larger ... ... *Rodneyanum*

Geranium.

Flowering stems, slender, elongate and leafy, from a thick root-stock; leaves on long stalks, orbicular in outline, deeply divided into 5 or 7 segments; petals small, pink; capsular valves hairy; seeds minutely reticulated... ... ... *pilosum*

Erodium.

Leaves 3-lobed; flowers few, petals blue ... ... *cygnorum*

Oxalis.

Dwarf; peduncles axillary few-flowered; petals small yellow; leaflets broadly obcordate; fruit cylindrical, pointed ... ... *corniculata*

ORDER SAPINDACEAE.

Stamens 8, turned to one side; sepals 5, petals 4 (the fifth wanting); ovules 2 in each cell. Diffuse shrub **Diplopeltis**

Stamens regularly arranged.
 Ovule 1 in each cell.
  Fruit separating into 3 distinct fruitlets each with a long terminal wing; petals 5 ... **Atalaya**
  Fruit of 4 hard lobes, only 1 or 2 developed, scarcely bursting; seeds half-enclosed in a crimson arillus; petals none ... ... **Heterodendron**
 Ovules 2 in each cell; *flowers unisexual*.
  Fruitlets 3, each with or without lateral wings; petals 0; stamens usually 8. Usually viscid shrubs ... ... **Dodonaea**

Diplopeltis.

Glandular-pubescent; leaves linear entire or 3-lobed ... *Stuartii*

Atalaya.

Small tree; leaves pinnate, glabrous; petals pubescent with a hirsute scale at the base ... ... *hemiglauca*

Heterodendron.

Small tree; leaves firm, entire, lanceolate, silky; flowers few, small in a short terminal panicle ... *oleifolium*

Dodonaea.

1. Leaves simple; each valve of the capsule produced into a vertical wing.
Leaves flat, lobeless, narrowed at the summit.
 Branchlets rounded; seeds of a dull lustre.

Leaves from elliptical to broad-linear, 3 or 4in., narrowed into a short stalk. A tall more or less viscid shrub; sepals usually 4; fruit 3-celled, the wings about as broad as the cells *viscosa*

Leaves oval-oblong, rounded at the base on rather long stalks ... ... *petiolaris*

Branchlets angular; seeds smooth shining; leaves lanceolate almost veinless; a tall shrub ... ... *lanceolata*

Leaves cuneate, mostly toothed or lobed atop. A low diffuse shrub, fruit 2- or 3-celled, wings narrow and thin ... ... *procumbens*

Leaves linear or linear-cuneate, serrately crenate or pinnatifid; fruit 3-celled; seeds shining. Erect shrub *lobulata*

II. Leaves simple; capsule not winged.

Leaves flat, cuneate or obovate, rigid.
  Sepals lanceolate; leaves small obovate; fruit 3- or 4-celled, dissepiments falling off with the valves. A small erect shrub ... ... *bursarifolia*

  Sepals broadly ovate; leaves small roundish or ovate, slightly sinuate-toothed; fruit 4- or 5-celled, the angles rarely produced into very narrow wings. A small erect shrub ... ... *Baueri*

Leaves linear, margins revolute, about 1in.; flowers solitary or 2 together on very short recurved stalks; stamens usually 6. A low glabrous shrub... ... *hexandra*

III. Leaves pinnate.

Capsule not winged.
  Leaflets 5 to 13, obovate-cuneate, deeply toothed at the end; flowers in short dense terminal corymbs; capsular globular, glandular-hairy ... *humilis*

Capsule winged as in *D. viscosa*.
  Leaflets obovate or cuneate-oblong, toothed at the end, rhachis dilated between the joints. Staminate flowers in small clusters on very short pedicels, pistillate ones 3 or 2 together or solitary. A low shrub, usually pubescent or hairy, very viscid ... ... *boronifolia*

  Leaflets linear; flowers clustered or racemose ... *tenuifolia*

  Leaflets linear, channelled; rhachis not dilated; fruits solitary stalked ... ... *stenozyga*

  Leaflets very short, oblong, obtuse; rhachis dilated; fruits 4-winged, solitary on rather long stalks *microzyga*

GENERA AND SPECIES.

ORDER STACKHOUSIEAE.

Petals 5, perigynous with free elongated claws but united upwards in a tubular corolla with spreading lobes. Erect herbaceous stalks emitted annually from a perennial root-stock; flowers in spikes; fruitlets usually 3 ... ... *Stackhousia*
Petals 5, hypogynous, free; fruitlets 5 ... ... *Macgregoria*

Stackhousia.

I. Corolla-lobes oblong-obtuse.

Fruitlets winged; leaves oblong-linear; spikes interrupted *megaloptera*
Fruitlets with 3 dorsal membranous angles; leaves obovate, fleshy; flowers crowded. Maritime ... ... *spathulata*
Fruitlets blunt, ovoid with reticulate markings; leaves lanceolate to linear.
  Spike dense at the top, usually interrupted as the flowering advances; petals pale-yellow ... ... *linarifolia*
  Spike short and dense; petals bright-yellow; leaves broadly linear ... ... *flava*
  Spike filiform, flowers distant; leaves narrow ... *muricata*

II. Corolla-lobes acute.

Fruitlets blunt, obovoid, reticulate-marked; spikes long and slender, flowers distant; leaves broad-linear ... *viminea*

Macgregoria.

Small erect herb; leaves acute; racemes terminal ... *racemigera*

ORDER PHYTOLACCEAE.

Fruit of 2 to 8 fruitlets; flowers axillary, calyx 4-cleft; stamens 6 or more; fruitlets bursting along outer edge ... ... *Didymotheca*
Fruit of many fruitlets; calyx sinuate-toothed.
  Flowers axillary, stamens 30-50 ... ... *Gyrostemon*
  Flowers in racemes; fruitlets bursting along inner edge. Small trees ... ... *Codonocarpus*

Didymotheca.

Slender herb; pistils 2, leaves linear-terete or filiform *thesioides*
Somewhat shrubby; pistils 3 to 8; leaves linear, channelled with a recurved point ... ... *pleiococca*

Gyrostemon.

Shrubby, fruitlets 15 to 20 bursting along inner or outer edge ... ... *ramulosus*

Codonocarpus.

Leaves narrow-linear; pistils 30-40, styles long. A small tree with slender stem ... ... *pyramidalis*

Leaves lanceolate or obovate; pistils 20-40, styles short. A small somewhat shrubby tree ... ... *cotinifolius*

ORDER MALVACEAE.

I. Floral bracts wanting.

Ovule solitary in each cell.
  Stigmas decurrent; flowers unisexual; fruitlets 5 or less, separating from the axis, irregularly bursting or indehiscent ... ... **Plagianthus**

  Stigmas terminal; flowers bisexual; fruitlets 5 or more, separating from the axis, imperfectly bursting or indehiscent ... ... **Sida**

Ovules 2 or more in each cell; stigmas terminal.
  Capsule consisting of 5 to 20 fruitlets, united at the base, each opening in 2 valves ... ... **Abutilon**

  Capsule 3-celled, opening in 3 valves ... ... **Howittia**

II. Floral bracts 3 or more.

Ovule solitary in each cell. Staminal column bearing filaments to the summit.
  Floral bracts 3, united at the base; stigmas decurrent; fruitlets many in a depressed circle round the prominent axis, indehiscent ... ... **Lavatera**

  Floral bracts 3, filiform, distinct; stigmas terminal; fruitlets 8-12, separating from the short axis, indehiscent or slightly 2-valved ... ... **Malvastrum**

Ovules more than one in each cell. Staminal column bearing filaments on the outside below the top, rarely to the top.
  Floral bracts 5 or more; style branched at the top, stigmas generally terminal; capsule 5-valved ... **Hibiscus**

  Floral bracts 3; style undivided, stigmas decurrent; capsule 3- to 5-valved ... ... **Gossypium**

Plagianthus

Flowers large white; leaves flat, membranous. Shrub *Berthae*
Flowers small, yellow; leaves fleshy.
  Flowers in dense terminal leafy spikes. Tall, erect, somewhat succulent herb; leaves glabrous ovate or ovate-oblong, on long stalks; calyx 5-angled, petals scarcely longer ... ... *spicatus*

Flowers axillary.
> Herb, low, much branched; leaves cuneate-oblong toothed at the end, slightly hoary; flowers in distant clusters along the leafy branches... *glomeratus*

> Shrub, dwarf, rigid, covered with scurfy scales; branches often spinescent; leaves linear to oblong-cuneate, very small, 3-toothed at the base; flowers minute, 1 to 3 together, almost sessile ... *microphyllus*

Sida.

1. Calyx not prominently ribbed. Carpels strongly reticulate on the sides. Petals yellow or whitish.

Calyx-lobes obtuse, not protruding beyond the fruit.
> Carpels 6 to 10, wrinkled on the back, glabrous; fruit 2½ to 4 lines diam. Semiherbaceous, procumbent, stellately hairy; leaves linear-elliptical to orbicular-cordate, crenate; stipules linear-filiform; flowers axillary, usually solitary, on slender stalks rarely as long as the leaves; petals yellow twice as long as calyx ... *corrugata*

> Carpels 5 to 8, hairy, but without wrinkles on the back; fruit 2 lines in diameter. A dwarf much-branched shrub, stellately hairy; with very small leaves and flowers ... *intricata*

Calyx-lobes acute, persistently herbaceous.
> Leaves lanceolate or oblong-linear; calyx very woolly; carpels 6 to 8, wrinkled on the back. An erect shrub with long twiggy branchlets ... *virgata*

> Leaves ovate- or orbicular-cordate; stamens few; carpels 1 line in diameter ... *cardiophylla*

Calyx-lobes acuminate with long subulate woolly points.
> Leaves ovate-lanceolate or cordate ... *cryphiopetala*

Calyx-lobes enlarged and thinner after flowering.
> Calyx-lobes ovate-lanceolate, of rather thick consistence. An erect shrub, beset with stellate tomentum; leaves lanceolate or oblong-linear, 1 to 1½ in. long, shortly stalked. Flowers 1 to 3 together on stalks shorter than the leaves; petals yellow, longer than the calyx; stamens 10-15; fruit depressed, tomentose, wrinkled on the back and furrowed between the carpels (usually about 7). ... *petrophila*

Calyx-lobes broadly ovate thin and transparent; habit, foliage, and inflorescence of *S. petrophila*, but the flowers larger and the fruit nearly globular ... ... *calyxhymenia*

II. Calyx 5-angled and 10-ribbed; carpels 10, not reticulate on the sides. Petals yellow.

A somewhat tall erect or spreading undershrub; leaves ovate or narrow, toothed, shortly stalked, nearly glabrous above, whitish with a short tomentum underneath ... ... *rhombifolia*

III. Fruiting-calyx with 15 to 20 prominent nerves. Carpels numerous. Leaves more or less orbicular. Undershrubs, densely velvety tomentose.

Fruiting calyx very large membranous, quite closed over the fruit ... ... *inclusa*

Fruiting calyx enlarging little after flowering, open at the top.
  Leaves ovate-cordate, 1 inch; petals broad, shorter than calyx; carpels about 24 ... ... *platycalyx*
  Leaves orbicular, about 1 inch; petals purple, glabrous, twice as long as calyx ... ... *lepida*

Howittia.

A tall erect shrub clothed with a rough stellate tomentum; leaves shortly stalked, ovate-lanceolate, 1 to 2 in.; flowers axillary, solitary; petals purple; style-branches very short ... ... *trilocularis*

Abutilon.

I. Capsule truncated or concave atop; fruitlets usually 2- or 3-seeded, angular-pointed or awned at the upper outer edge.

Carpels 10 or less, not exceeding the calyx-lobes, the points usually erect. Stems shrubby.
  Calyx-lobes shorter than the tube.
    Petals adnate high up the glabrous staminal tube; calyx tubular, 1 inch long ... ... *tubulosum*
    Petals shortly adnate to the pubescent base of of the staminal column; calyx campanulate.
      Petals white, 1 inch; more than twice as long as calyx ... ... *leucopetalum*
      Petals yellow, shortly exceeding the calyx *Mitchelli*
      Petals very small or shortly exceeding the rather inflated calyx ... ... *cryptopetalum*

GENERA AND SPECIES. 33

 Calyx-lobes longer than the tube, very concave strongly keeled and acuminate; fruitlets about 10, ear-shaped, much compressed, rather obtuse or scarcely pointed ... ... *otocarpum*
 Carpels 10 to 15, exceeding the calyx-lobes, the points divergent. Stems herbaceous.
  Stem coarse and erect; leaves broadly cordate, 3 to 4 in., carpels 10 to 15 with long divergent points ... ... *Avicennae*
  Stem slender; leaves ovate or cordate-lanceolate, 1 to 3 in., carpels about 10 with short divergent points ... ... *oxycarpum*
   II. Fruitlets usually 1-seeded, rounded or angled at the upper outer edge.
 Leaves cordate, crenate, about 1 in.; flowers $\frac{1}{2}$ in. or more in diameter much exceeding the calyx. Capsule exceeding the calyx, depressed in the centre, slightly tomentose or pubescent ... ... *Fraseri*
 Leaves orbicular, truncate or retuse ... ... *halophilum*
 Leaves ovate-oblong under 1 in.; flowers minute; carpels 3-seeded ... ... *macrum*

Lavatera.

A coarse erect, branched, somewhat shrubby plant; leaves orbicular-cordate, 5- to 7-lobed, on long stalks; flowers large, pink or white, usually several together, on short stalks, in the axils of the leaves   *plebeia*

Malvastrum.

An erect branching herb of 1 to 2 feet; flowers in short terminal spikes; leaves stalked ovate or ovate-lanceolate ... ... *spicatum*

Hibiscus.

I. Floral bracts free.

Calyx shortly 5-lobed, inflated, enclosing the hairy capsule; seeds glabrous. Erect annual, leaves deeply 3- or 5-lobed ... ... *trionum*
Calyx deeply 5-lobed.
 Seeds covered with woolly hairs.
  Staminal tube short with long filaments round the summit; lower leaves small, orbicular... *brachysiphonius*
  Staminal tube slender, the filaments not extending beyond the middle.
   Leaves ovate or lanceolate, entire ... *microchlaenus*
   Leaves orbicular, broadly 3-lobed ... *Pinonianus*

C

Seeds shortly pubescent. Small velvety-tomentose shrub; leaves ovate or ovate-lanceolate, toothed; floral-bracts about 7, linear-subulate almost free, shorter than the calyx ... ... *Krichauffi*

II. Floral bracts united, at least at the base.

Leaves undivided.
Leaves cordate-ovate, 1 to 1½ in. long; involucre of floral bracts with 7 or 8 short lobes; seeds glabrous ... ... *Sturtii*
Leaves cordate-orbicular, 4 to 5 in. long; involucre with 8 to 10 obtuse lobes; seeds tomentose ... *Farragei*

Leaves lobed; capsule tomentose.
Involucre with 7 to 10 linear teeth; calyx tomentose. A tall shrub with a scabrous tomentum; leaves deeply 3- to 5-lobed; flowers very large, bluish-purple; seeds glabrous ... ... *Wrayae*
Involucre with 3 to 6 short rigid teeth; calyx glabrous, black-dotted. A tall glabrous shrub; leaves from deeply bipinnatifid to trifid, upper leaves entire; flowers large; seeds woolly ... *hakeaefolius*

Gossypium.

Floral bracts 3, linear; leaves ovate, more or less sinuate or 3-lobed. An undershrub, hoary with a dense short tomentum, flowers large pink ... ... *australis*

Floral bracts 3, cordate; leaves broadly ovate entire. A shrub of a few feet, glabrous and marked with black dots; flowers very large, purple with a dark centre, on short stalks in the upper leaf-axils ... *Sturtii*

ORDER TILIACEAE.

Fruit globular, prickly, not bursting; petals narrow hairy or with a pit at the base ... ... **Triumfetta**
Fruit long, smooth, valvular; petals ovate or broad ... **Corchorus**

Triumfetta.

Ovary 3-celled; fruit ovoid-globular, thinly tomentose, prickles long, slenderly hooked; leaves oval, velvety... *Winneckeana*

Corchorus.

Capsule slender, long, tomentose. An erect tomentose shrub with small flowers in nearly sessile clusters ... *sidoides*
Capsule ovate-globular, short, stellately hairy ... *Elderi*

ORDER STERCULIACEAE.

I. Petals flat longer than the calyx.

Stamens 5, free, opposite the petals; anthers with bifid apices; ovary 5-celled; embryo curved ... ... **Hermannia**

Stamens 5, united at the base, no staminodia; ovary 1-celled ... ... **Waltheria**

Stamens 5, united into a cup, 5 staminodia; ovary 5-celled ... ... **Melhania**

II. Petals dilated and inflexed at the base, narrowed at the summit.

Stamens 5, scarcely united at the base, with intervening staminodia; fruit 5-valved ... ... **Commergonia**

III. Petals wanting; or small and scale-like, shorter than the calyx.

Flowers unisexual, petals 0, stamens 15, inserted on a column; fruit of distinct follicular fruitlets; floral bracts 0... ... ... **Brachychiton**

Flowers bisexual; stamens 5, shortly connate, alternate with sepals.

   Anthers opening outwards by parallel slits.

      Capsule membranous; calyx enlarged after flowering, thin, coloured; petals and floral bracts none ... ... **Seringia**

      Capsule woody; calyx strongly ribbed; petals 5; floral bracts 1-3 ... ... **Hannafordia**

   Anthers opening in terminal pores; petals 0, or very small; floral bracts 3; stamens 5.

      Calyx divided to the middle, enlarged and coloured after flowering, prominently ribbed; stipules present... ... ... **Thomasia**

      Calyx divided almost to the base, scarcely enlarging, many-veined at the base; stipules none. ... ... **Lasiopetalum**

Hermannia.

Herb; leaves ovate, crenate, stalked; flowers blue solitary ... ... *Gilesii*

Waltheria.

A small undershrub; flowers small yellow in dense heads ... ... *Indica*

Melhania.

A slender velvety-tomentose shrub; petals large yellow ... ... *incana*

Commersonia.

Leaves oblong-lanceolate, serrate, densely woolly; flowers large; calyx-bud rayed at the summit ... *magniflora*

Leaves obliquely ovate-cordate, serrate, stellately hairy; flowers small; calyx-bud scarcely angled... *loxophylla*

Leaves narrow-ovate, irregularly serrate, densely stellate-tomentose, rather small; flowers small ... *Kempeana*

Leaves linear-oblong or spathulate, very small, margins recurved, shortly lobed towards and at the summit, scabrous hairy; flowers very small, in a few-flowered cyme. Slender diffuse undershrub ... *Tatei*

Brachychiton.

Shrub with lobed leaves; calyx bell-shaped ... *Gregorii*

Seringia.

Bracts narrow; carpels several-seeded, seeds ovoid; leaves oblong-lanceolate, rugose and pubescent ... *corollata*

Bracts broad, scarious, coloured, deciduous; carpels 1- or 2-seeded, seeds reniform.
 Leaves thick and soft 1 to 2 inches long ... *nephrosperma*
 Leaves smooth under 1 inch long ... *integrifolia*

Hannafordia.

Velvety hispid; leaves oblong-lanceolate, entire; calyx-teeth subulate-linear ... *Bissilii*

Thomasia.

A small erect, stellately hairy shrub; leaves oblong-oval, wrinkled; stipules large leafy semihastate; flowers few, large, in racemes; calyx lilac; petals scale-like, barren stamens subulate; fruit 3-celled ... *petalocalyx*

Lasiopetalum.

1. Style glabrous; erect shrubs with stiff leaves, white or rusty-tomentose underneath; calyx pink or red.

Bracts longer than the calyx, petal-like, forming an involucre round the soft woolly flower-heads; leaves cordate-ovate, stalked; calyx glabrous inside ... *discolor*

Bracts not exceeding the calyx; leaves shortly stalked Calyx glabrous inside; floral bracts subulate, mostly narrow-elliptical ... *Behrii*

Calyx tomentose inside.
>> Leaves linear or oblong-linear; floral bracts small, lanceolate; flowers small, few, on slender stalklets, forming a short racemose cyme ... *Baueri*
>> Leaves broader; floral bracts petaloid, flowers larger, several, forming cymes ... *Tepperi*

> II. Style densely covered with stellate hairs.

An erect shrub with large flaccid cordate leaves; flowers large in loose cymes; floral bracts 1 or 2, small, linear-filiform; sepals white, glabrous inside ... *Schulzeni*

ORDER TREMANDREAE.

Tetratheca.

Anthers continuous with the filament, 4-celled, 2 in front of the 2 others; seeds hairy with an appendage at the chalazal end. Heath-like shrubs with round stems and rather large pink flowers.

> Leaves broadly ovate, clustered in 3's or 4's. Ovules 2 superimposed in each cell. ... *ciliata*
> Leaves linear, scattered, rarely wanting or reduced to scales; ovule 1 in each cell; flowers rarely white ... *ericifolia*

ORDER EUPHORBIACEAE.

> I. Flowers without sepals and petals included within a calyx-like or petaloid involucre.

Flower-clusters consisting of one pistillate flower surrounded by several staminate flowers each of one stamen on a pedicel, intermixed with scales; the calycine involucre with glands on its margin; fruit 6-valved. Herbs with a milky acrid juice ... **Euphorbia**

> II. Flowers with sepals. No involucre.

Stamens 10 or less.
> Anthers opening by terminal pores.
>> Sepals 5, white; petals minute; stamens 5, free **Poranthera**
> Anthers opening by longitudinal slits.
>> Petals absent.
>>> Leaves in alternate clusters of 3; sepals 6, petal-like; stamens 3 to 9, free ... **Micrantheum**
>>> Leaves scattered.
>>>> Fruit 1-celled, 1-seeded; stamens 3, free; styles 3, undivided; sepals 6 **Pseudanthus**

Fruit 3- or more-celled.
   Stamens 3, free or united; styles 3, free or united, usually bilobed; sepals 6... ... ... **Phyllanthus**
   Leaves absent. Stamens 10 or fewer, free; *calyx 3- to 5-lobed;* style-branches entire; *capsule with 6 erect tooth-like appendages* **Amperea**
   Petals present. Style-branches fringed; flowers in terminal cymes. Stamens free ... ... **Monotaxis**
Stamens numerous. Petals present.
   Petals shorter than the sepals; flowers small, axillary; stigma sessile, entire or scarcely lobed; stamens free ... ... **Beyeria**
   Petals longer than the sepals; flowers in terminal clusters; styles 3, deeply bifid; stamens united **Ricinocarpus**
Stamens numerous. Petals absent.
   Flowers axillary, solitary or few together; styles 3, divided into 2 to 4 branches; stamens united **Bertya**
   Flowers in short few-flowered terminal spikes; *styles 3, bifid, beset with papillae;* stamens free; sexes in different plants ... ... **Adriana**

Euphorbia.

I. Dwarf diffuse or prostrate.

Stem and leaves downy; leaves roundish- or cuneate-ovate, serrulated; involucral glands denticulated, bordered by a red lobe; seeds without appendage *erythrantha*

Stem and leaves glabrous; leaves obliquely elliptical to orbicular; involucral glands entire with a narrow white border; seeds without appendage ... ... *Drummondi*

II. Stems erect glabrous.

Leaves oval-oblong, oblique; involucral glands with a petaloid appendage; seeds deeply rugose ... ... *Wheeleri*

Leaves linear or linear-lanceolate; involucral glands broad, reniform, brownish, undivided, without any appendage; seeds granular-rugose *with a large appendage* ... ... *eremophila*

Poranthera.

Dwarf, diffuse, glabrous, annual; leaves flat, linear to obovate, obtuse, stalked, spathulate; flowers very small collected in short leafy corymbs ... ... *microphylla*

Dwarf, erect, glabrous perennial; leaves linear with revolute margins, crowded, sessile; flowers forming a broad compact corymb ... ... *ericoides*

Micrantheum.

A heath-like shrub, glabrous or the branchlets slightly pubescent; leaves rigid, linear or oblong; flowers 1 or few-together; axillary; stamens 6 to 9 ... ... *hexandrum*

Pseudanthus.

A rigid much-branched glabrous shrub; leaves mostly ovate, 1 or 2 lines long, very shortly stalked ... *micranthus*

Phyllanthus.

I. Stamens with connate anthers; styles distinct or united.

Sepals of male flowers, narrow, erect, connate in a more or less tubular calyx.
  Herbaceous, glabrous; leaves from narrow- to cuneate-linear or the lower ones spatular-cuneate, flat or recurved at the margin; styles united ... *thesioides*
  An intricately branched spinescent shrub; leaves very small obcordate or cuneate, clustered; styles 3, very short spreading; male flowers sessile, female flowers on long pedicels ... ... *rigens*

Sepals of male flowers ovate, spreading.
  A somewhat tall shrub, glabrous; leaves oblong or broadly linear, arranged in 2 rows; stipules subulate; styles short, very shortly lobed ... ... *rhytidospermus*
  A dwarf shrub with pubescent branchlets; leaves oval or elliptical, scattered ... ... *Tatei*

II. Stamens with the anthers and filaments free; styles free.

Fruiting-calyx as long or longer than the capsule.
  A glabrous shrub; leaves oblong-cuneate, very obtuse, scarcely stalked; flowers on rather long stalks; capsule glabrous; seeds ribbed, striate or smooth ... ... *calycinus*
  A hoary undershrub; leaves spatular or elliptic-obovate, shortly stalked; flowers on short stalks; capsule pubescent; seeds smooth... ... ... *Fuernrohrii*

Fruiting-calyx shorter than the capsule.
  Low diffuse glabrous annuals.
    Leaves elliptical, in two rows, almost sessile; stipules very minute spreading; fruit attenuated upwards; seeds large rough ... *trachyspermus*
    Leaves oblong or linear-cuneate very shortly stalked; stipules minute; fruit depressed-globular; seeds finely striated ... ... *lacunarius*

Undershrubs with small, stiff, scattered leaves.
    Low diffuse glabrous undershrub; leaves ovate
    or obovate; capsule very small ... ... *australis*
    Erect dwarf shrub, with hairy twiggy branch-
    lets; leaves cuneate- or elliptic-ovate, re-
    curved at the margin; stipules small, black;
    capsule small ... ... *thymoides*
Tall shrub, glabrous; leaves large, membranous,
    obovate or orbicular, in 2 rows; flowers in axil-
    lary clusters on conspicuous pedicels; capsule
    small ... ... *Gunnii*

Amperea.

A perennial herb with erect, rigid, triangular, leafless
    stems; stipules small, deeply fringed or lobed ... *spartioides*

Monotaxis.

Leaves large, oblong to narrow-lanceolate, flat; petals
    yellow, very obtuse ... ... *luteiflora*

Beyeria.

A tall shrub, the branchlets viscid; leaves from oval-
    oblong to broadly linear, 1 to 2 in. long, shining
    above, white tomentose underneath; flowers 2 or
    3 together, on recurved pedicels much longer than
    the calyx; capsule about 4 lines long ... ... *viscosa*
A small erect shrub; leaves broad-linear, blunt, hardly
    viscid, under 1 inch; pedicels about as long as the
    calyx; capsule smaller... ... ... *opaca*
An erect viscid shrub; leaves narrow-linear, recurved
    or hooked at the point ... ... *uncinata*

Ricinocarpus.

An erect glabrous shrub of 2 to 3 ft.; leaves rigid,
    linear; flowers sweet-smelling, on long pedicels;
    petals white, about ½ in. long, usually 6; fruit
    nearly globular, about ½ in. in diameter ... ... *pinifolius*

Bertya.

An erect shrub, the branchlets and foliage stellately
    tomentose; leaves narrow with recurved margins.
    Ovary densely villous; flowers solitary almost ses-
    sile; calyx-segments rather broad, nearly 2 lines
    long ... ... *Mitchell*

An erect shrub or small tree, the branchlets stellately tomentose; leaves ovate, very convex, glabrous above, white-tomentose underneath; ovary stellately tomentose ... ... *rotundifolia*

Adriana.

An erect shrub, more or less beset with stellate hairs; leaves opposite almost sessile, ovate-lanceolate or oblong, coarsely toothed, about 2 in. long, with 2 stalked glands at the base ... ... *quadripartita*

A small shrub; leaves conspicuously stalked, alternate, 3-lobed ... ... *tomentosa*

ORDER PORTULACEAE.

Petals (yellow) and stamens perigynous; ovary half-inferior. Prostrate or spreading succulent herbs ... **Portulaca**

Petals and stamens hypogynous; ovary superior. Branches usually slender, erect. Succulent herbs ... **Claytonia**

Portulaca.

Leaves mostly alternate.
    Stipular hairs very minute. Leaves oblong-cuneate; flowers terminal; petals usually scarcely longer than the calyx, slightly united at the base... ... *oleracea*
    Stipular hairs numerous and conspicuous.
        Leaves thick, oblong under ½ inch long ... ... *australis*
        Leaves linear-terete, longer; flowers larger ... *filifolia*
Leaves opposite, orbicular, no stipular hairs, petals minute ... ... *bicolor*

Claytonia.

I. Stamens 20 to 100.

Petals 8 or 9; style 4-cleft; seeds minute... ... ... *pleiopetala*

Petals 5, broad, purple; style 3-lobed; seeds large, black, reticulated. Leaves thick, fleshy, oblong-spathulate; flowers large in loose terminal racemes on long pedicels *Balonnensis*

Petals 5, narrow-ovate, large; styles 3; seeds pitted ... *polyandra*

II. Stamens usually less than 10; capsule ovoid or oblong, bursting.

Stamens 8 to 10; seeds pitted; sepals broad obtuse. Stems ascending short or much elongated; leaves crowded on lower part of stem, linear-oblong; flowers pink, few in loose racemes ... ... *volubilis*

Stamens not many; seeds longitudinally furrowed; leaves linear-filiform ... ... *ptychosperma*

Stamens 3 to 5 ; seeds smooth, shining.
 Leaves oblong or linear-oblong, thick.
  Bracts foliaceous, pedicels short ... ... *brevipedata*
  Bracts small, scarious.
   Stems ascending, much branched, from 1 to few inches long ; radical leaves stalked, linear-oblong or spathulate ; stem-leaves few, smaller ; sepals acute, lengthening after flowering ; capsule longer than calyx *calyptrata*
   Leaves all radical ; sepals obtuse ... ... *pumila*
 Leaves narrow-linear.
  Flowers very small ; stamens usually 3 ; capsule narrow-cylindrical ; stems numerous, ascending to a few inches, leaves chiefly radical and stalked. Racemes numerous, short, axillary and terminal ... ... *corrigiolacea*
  Flowers large white ; stamens 5, opposite to and adherent to the base of the petals ; capsule ovoid. Plant tufted with a creeping stem ; racemes few-flowered, terminal or leaf opposed *Australasica*
 III. Stamens not exceeding 10 ; capsule globular scarcely bursting.
  A minute annual with decumbent or ascending stems ; leaves oblong to ovate, thick ; petals 5 to 7 a little longer than the calyx ; stamens 7 to 10 united at the base ; seeds small, minutely pitted ... ... *pygmaea*

ORDER CARYOPHYLLEAE.

Sepals united, no stipules ; styles 2.
 Petals 5, capsule many-seeded, calyx tubular ... **Saponaria**
Sepals free ; no stipules.
 Styles 3, capsular valves 6 ; petals bifid ... ... **Stellaria**
 Style 3-cleft, capsular valves 3, petals lobed... ... **Drymaria**
 Styles and capsular valves 4 to 5.
  Stamens 4 to 5, opposite to sepals ... **Sagina**
  Stamens 4 to 5, alternate ... ... **Colobanthus**
Sepals free ; stipules present.
 Styles 3, distinct from the base **Spergularia**
 Styles 3, united at the base ... ... **Polycarpon**
 Style long, 3-toothed ... ... **Polycarpaea**

Saponaria.

A slender erect dichotomous annual beset with sticky hairs, leaves very small filiform ... ... *tubulosa*

Stellaria.

Petals white, perennial herbs, pedicels axillary usually
longer than the leaves.
   Much branched, decumbent; leaves rigid, pungent-
   pointed, linear-lanceolate, often recurved ... *pungens*
   Branches usually slender, erect; leaves linear, slender; or rarely densely tufted ... ... *glauca*
Petals none; a slender, glabrous, branching annual;
leaves mostly lanceolate; pedicels axillary shorter
or longer than the calyx... ... ... *multiflora*

Drymaria.

A glabrous dichotomously branched annual; leaves chiefly
crowded at the base, narrow-linear; pedicels slender
about ½ inch long, axillary; capsule cylindrical ... *filiformis*

Sagina.

A small glabrous annual, somewhat tufted with ascending branches; pedicels erect longer than the leaves,
petals minute or wanting; leaves small, subulate,
joined by a scarious sheath ... ... *apetala*

Colobanthus.

A small densely tufted, stemless perennial; leaves linear-
subulate, sometimes very rigid. Peduncles 1-flowered
shorter or longer than the leaves ... ... *Billardieri*

Spergularia.

A small low spreading annual beset with short viscid
hairs; leaves narrow-linear; flowers pink on short
pedicels in forked cymes; capsule about as long as
calyx; seeds not winged ... ... *rubra*
A glabrous perennial, with long ascending branches,
flowers large white; capsule very large, twice the
length of the calyx; seeds surrounded by a broad
scarious wing ... ... *marina*

Polycarpon.

A glabrous prostrate annual; leaves obovate or oblong,
clustered in fours; flowers very small, numerous in
loose terminal cymes ... ... *tetraphyllum*

Polycarpaea.

A glabrous annual, stems erect or decumbent from a
rosette of oblong leaves; petals united below with
the stamens; flowers in terminal cymes ... ... *synandra*
Branches erect slender, minutely pubescent; leaves
narrow-linear; petals quite free ... ... *Indica*

ORDER ILLECEBRACEAE.

Stipules present; bracts scarious, whorled; stigmas 2, sepals 5, petals 5, filiform	**Herniaria**
Sepals connate; leaves connate at the base; styles 2 ...	**Scleranthus**

Herniaria.

Minute stems, calyces and margin of leaves beset with short white rigid hairs; flowers in axillary clusters; leaves narrow-elliptical	*incana*

Scleranthus.

Low tufted perennial, leaves rigid, sharp-pointed; flowers in sessile clusters; stamens 5, with staminodia; calyx-lobes broad, scarious	*pungens*
Calyx-lobes narrow, scarious; stamens 2	*diander*

ORDER POLYGONACEAE.

Sepals 6, the 3 inner ones larger, callously thickened, spreading over the fruit, the 3 outer ones spreading; stamens 6; stigmas bearded. Herbs ...	**Rumex**
Sepals 5, nearly equal; stamens 5 to 8. Styles 2 or 3; flowers bisexual. Herbs	**Polygonum**
Calyx *enlarged* and often succulent *in age*; flowers unisexual. Shrubs	**Muehlenbeckia**

Rumex.

I. Leaves chiefly radical.

Flower-clusters distant without floral leaves; leaves elongate, lanceolate-elliptic; inner sepals hooked at the tip, toothed at the base	*Brownii*

II. Floral leaves present, all longer than the flower-clusters.

Inner fruit-sepals rather large, without tubercles, straight-pointed, shortly toothed on the sides; panicle much divaricately branched	*flexuosus*
Inner fruit-sepals small, with a dorsal tubercle, and one or two very narrow lobes on each side; rather dwarf, leaves lanceolate crisped on the margin ...	*crystallinus*
Inner fruit-sepals rather large, rhomboid, with a faint dorsal tubercle, lobeless; male flowers clustered above, female flowers in the lower axils. Stem tall, simple, hollow	*bidens*

Polygonum.

I. Flowers in axillary clusters.

Stems prostrate, short compact; leaves very small, narrow-spathulate to linear; nut triangular smooth and shining ... ... *plebeium*

II. Flowers in spikes.

Prostrate, somewhat hairy; spikes axillary, solitary, shortly stalked; fruit triangular ... ... *prostratum*

Slender erect, glabrous; spikes more or less paniculated.
Sepals glandular-dotted, green or reddish
    Stipular-tube truncate without cilia; leaves with glandular dots ... ... *lapathifolium*

    Stipular-tube inflated, shortly ciliated ... *Hydropiper*

Sepals without glandular dots.
    Stipular-tube bordered by long cilia; sepals red; fruit bi- or tri-angular; leaves mostly narrow-lanceolate ... ... *minus*

    Stipular-tube shortly ciliate; leaves lanceolate tapering into a very long point, and contracted into a rather long stalk ... ... *attenuatum*

Muehlenbeckia.

Stem prostrate or climbing; leaves large, broad; flowers in interrupted spike-like racemes; fruit-calyx usually very succulent ... ... *adpressa*

Stems erect; leaves narrow, tapering at the base.
    Flowers in terminal spikes; branchlets slender, interlacing; fruit prominently 3-angled; leaves linear ... ... *Cunninghami*

    Flowers in axillary clusters; strictly erect; fruit globular, 3-angled; leaves broadly linear ... *polygonoides*

ORDER CHENOPODIACEAE.

I. Leaves developed. Embryo almost annular

Calyces of two forms;—calyx of staminate flowers, small equally lobed; calyx of pistillate flowers with 2 appressed segments enlarged in fruit ... **Atriplex**

Calyx of 3 or 1 minute dry clavate segments... ... **Dysphania**

Calyx equally 5-lobed, without appendages.
    Fruit a berry; seed flattened, horizontal ... ... **Rhagodia**

    Fruit dry, fruiting calyx unchanged; seed horizontal flattened, or vertical and less compressed **Chenopodium**

Fruiting calyx enlarged and succulent.
  Fruiting calyx depressed-globular, seed horizontal ... ... **Enchylaena**
  Fruiting calyx elliptical, seed vertical ... **Threlkeldia**
Calyx with appendages, closing over the fruit.
  Fruiting calyx enlarged with 5 free or connate horizontal wings. Undershrubs, rarely herbs ... **Kochia**
  Fruiting calyx indurated and mostly enlarged into 5 or less spinous appendages. Small undershrubs or almost shrubby ... ... **Bassia**
  Fruiting calyx with small membranous lobes and 2 or 3 dorsal stalked, more or less vertical, wing-like membranes. Small diffuse undershrubs; leaves fleshy, cylindrical ... ... **Babbagia**

  II. Leaves absent. Embryo almost annular.

Stems jointed, fleshy; flowers on each side of the shorter terminal joints of the branches forming a more or less compact terminal spike; calyx thin 2- to 5-lobed; stamens 1 or 2; styles 2 or 3, united in a column; fruit concealed within the joints ... **Salicornia**

  III. Leaves developed; embryo spirally coiled.

Calyx rigid with dorsal horizontal wings; leaves almost cylindrical, pungent-pointed... ... ... **Salsola**
Calyx herbaceous, without appendages; leaves almost cylindrical, thick .. .. **Suaeda**

Atriplex.

I. Staminate and pistillate flowers on separate plants. Fruiting calyx closed to near the base. Shrubs with the male inflorescence in terminal spikes, the female flowers generally solitary and axillary.

Fruiting calyx flat, reniform, on a stalk nearly as long as the calyx; leaves narrow-elliptical to obovate *stipitatum*

Fruiting calyx flat, ovate, inconspicuously stalked; leaves from elliptical to lanceolate ... ... *paludosum*

Fruiting calyx with thick convex valves; leaves orbicular to ovate-rhomboid ... ... *nummularium*

Fruiting calyx flat, deltoid to rhomboid, with a turbinate solid base; leaves lanceolate to oval, ashy-grey *cinereum*

  II. Staminate and pistillate flowers on the same plant.

Male flowers in terminal spikes; female flowers solitary and axillary. Shrubs.

GENERA AND SPECIES.

Fruiting calyx closed only near the base, each segment with a large inflated dorsal appendage; leaves elliptical or oblong-lanceolate ... ... *vesicarium*

Fruiting calyx closed to near the middle, without appendages; leaves hastate-ovate, entire, or somewhat toothed; fruit-calyx 1½ to 3 lines diameter ... ... *rhagodioides*

Fruit-calyx 4 to 5 lines diameter, with a narrow flat margin ... ... *incrassatum*

Male flowers in axillary clusters with a few females in the upper axils; female flowers clustered in the lower axils without males. Herbs.
Fruiting calyx open to the base.
   Fruiting calyx entire, scaly tomentose; leaves ovate or rhomboid, coarsely sinuate-toothed *velutinellum*

   Fruiting calyx with thick rhomboid segments, their margins laciniated, densely scaly tomentose ... ... *fissivalve*

   Fruiting calyx on long stalks; the segments renate, entire or distantly denticulated, with a small prominent renate appendage at their base. Somewhat succulent, beset with watery papillae... ... ... *Quinii*

Fruiting calyx closed only near the base, raised on a short stalk, segments large denticulated; leaves orbicular or broadly rhomboid on long stalks ... *angulatum*

Fruiting calyx closed to near the middle.
   Fruit-calyx succulent and red towards the base. Procumbent; leaves narrow, often toothed ... ... *semibaccatum*

   Fruit-calyx rhomboidal, turgid; an erect herb with rather large obovate or rhomboidal leaves, coarsely and irregularly sinuate-toothed or lobed ... ... *Muelleri*

   Fruit-calyx rhomboidal under 1 line broad. Dwarf, diffuse or procumbent; scaly tomentose; leaves ovate or lanceolate, very shortly stalked ... ... *prostratum*

Fruiting calyx closed to near the summit, not at all compressed, the orifice small and closed by small erect appressed valves.
   Fruit-calyx narrow cylindrical; calyx-lobes minute, entire ... ... *leptocarpum*

     Calyx-lobes with two pointed excrescences and a dorsal appendage between them *limbatum*

Fruit-calyx succulent and green. A prostrate succulent shore-plant, with small elliptical or rhomboid denticulated leaves covered with watery papillae ... ... *crystallinum*

Fruiting calyx inflated of a spongy texture minutely bilobed at the summit.
Fruit-calyx flat-topped and acute at the margin ... ... *halimoides*
Fruit-calyx rounded at the summit ... *holocarpum*

Dysphania.

Calyx consisting of 3 segments, Erect herbs of a few inches, the inflorescence in dense terminal spikes.
Leaves ovate, flat; calyx-segments obovate; fruit obovate ... ... *plantaginella*
Leaves ovate-oblong, wavy on the margin, on slender stalks; calyx-segments united at the base; fruit broadly ovate ... ... *simulans*

Calyx consisting of a single sepal. Leaves very small, oval to elliptical; flowers in axillary clusters. Small prostrate or ascending herb ... ... *litoralis*

Rhagodia.

I. Panicle usually much branched.

Leaves oblong-lanceolate, greenish. A straggling or erect shrub up to 6 ft.; fruit usually red ... ... *Billardieri*
Leaves oval, mealy white on both sides. Tall, erect... *parabolica*

II. Inflorescence a spike or simple panicle.

Leaves thick and fleshy, alternate, branchlets not thorny.
Leaves mostly hastate, broadly orbicular or deltoid; fruit-calyx exceeding the fruit. A divaricately branched shrub covered with a dense white tomentum ... ... *Gaudichaudiana*
Leaves linear-cuneate, shortly stalked, about ½in. long; fruit-calyx not exceeding the fruit. A diffuse shrub ... ... *crassifolia*
Leaves green, or scarcely hoary, about 1in. long; inflorescence usually more slender and elongated *Preissii*

Leaves flat, thin, mostly scattered, small and broad, somewhat mealy-white; branchlets often ending in thorns. A divaricately branched rigid shrub ... *spinescens*

Leaves flat, thin, mostly opposite, lanceolate or broad, acute; fruit red or yellow. A slender climbing or somewhat prostrate and diffuse herb ... ... *nutans*

Chenopodium.

Tall shrub, branchlets often spinescent. Leaves spathulate or linear-elliptical, entire; flowers in terminal spikes or panicles; seeds vertical ... *nitrariaceum*

Herbs, mealy-white; inflorescence terminal or axillary; seeds horizontal.
   Erect, tall; leaves stalked, ovate or oblong, entire about 1 in.; spikes terminal, yellow ... ... *auricomum*
   Prostrate, small; leaves small, ovate or broadly lanceolate; flower-clusters mostly axillary ... *microphyllum*

Herbs, more or less glandular-hairy, not mealy, prostrate; inflorescence axillary; seeds vertical.
   Sepals deeply concave; flowers in spikes; leaves small ovate-lanceolate, acutely toothed; stamens 1 or more ... ... *rhadinostachyum*
   Sepals bluntly keeled, broad, concave; leaves small on long stalks, ovate or oblong, wavy sinuous at the margin; stamen 1. A small annual ... *carinatum*
   Sepals linear, acute, fringed along the keel; otherwise like *C. carinatum* ... ... *cristatum*
   Sepals acute, much thickened at the base; leaves lanceolate to broadly hastate, entire, on long slender stalks; stamen 1 ... ... *atriplicinum*

Enchylaena.

Prostrate or diffuse, downy (rarely almost glabrous) undershrub; leaves linear; flowers sessile, axillary and solitary; stamens 5; calyx-fruit red or yellow *tomentosa*

Threlkeldia.

Prostrate or diffuse, glabrous undershrub; leaves succulent, linear; flowers sessile, axillary and solitary; stamens 5; calyx-fruit somewhat succulent, dark-violet ... ... *diffusa*

Kochia.

1. Fruit-calyx bordered by distinct horizontal wings.

Wings fringed; terminal lobes of calyx erect, broad and denticulated; leaves linear-acute ... ... *fimbriolata*
Wings invested with minute hairs, alternating with 5 linear-acute appendages; leaves rather long ... *lanosa*
Wings distinctly stalked, alternating with linear-spathulate reflexed appendages ... ... *lobiflora*

Fruit-calyx without any appendages besides the entire horizontal wings; leaves very short.
    Leaves thick, ovate-triangular, mostly opposite   *oppositifolia*
    Leaves almost terete, scattered   ...   ...   *brevifolia*

    II. Fruit-calyx bordered by membranous horizontal wings, more or less united.

Fruit-calyx with vertical wings.
    Vertical wings 3; horizontal wings united. Erect, glabrous; leaves linear semiterete   ...   ...   *triptera*
    Vertical wings 5; horizontal expansion with 5 subovate lobes, glabrous above; fruit spongy ...   *decaptera*
    Vertical wings 5; horizontal expansion almost complete, slightly lobed. Erect herbaceous stems clothed with dense cottony wool; fruit hard ... ...   *pentatropis*

Fruit-calyx without vertical wings.
    Fruit-calyx pyramidal atop. Shrubby, erect, shortly hairy; leaves short, linear or terete, thick   *pyramidata*
    Fruit-calyx flat atop or nearly so, enveloped in dense wool ... ...   *eriantha*
    Fruit-calyx flat atop or nearly so, glabrous or tomentose.
        Leaves linear-cylindric, about ½in.; horizontal expansion usually red. Shrubby, more or less hairy ... ...   *villosa*
        Leaves crowded, short, thick, velvety-downy   *sedifolia*
        Leaves minute or none; branches wiry, ridged, almost glabrous, spinescent ... ...   *aphylla*
        Leaves flat, lanceolate-linear, comparatively long. Dwarf herb ... ...   *humillima*

    III. Fruit-calyx bordered by an undivided rigid annular expansion.

Expansion fringed with long soft hairs ... ...   *ciliata*
Expansion 5-angled; tubular part of calyx 5-ribbed ...   *brachyptera*
Expansion with 10 to 12 short, rigid, radiating points   *stelligera*

Bassia.

    I. Fruit-calyx without spines (also *B. biflora*). Flowers solitary, axillary.

Fruit-calyx small, obliquely globose, indurated, glabrous, with an obtuse hollow protuberance on one side. Prostrate, nearly glabrous; leaves narrow-linear; seed vertical ... ...   *salsuginosa*

Fruit-calyx not indurated, almost glabrous, with 5 unequal, dilated, and reflexed lobes. A dwarf hairy herb; leaves lanceolate; seed horizontal ... *enchylaenoides*

Fruit-calyx globose, not indurated, with 5 membranous lobes. An undershrub clothed with cottony wool; seed horizontal ... ... *Dallachyana*

Fruit-calyx depressed, globular, of thin texture, densely tomentose, with 3 long radiating soft woolly horns, obtuse and turned up at the end ... ... *tricornis*

II. Fruit-calyx much indurated, with 1 or 2 spines. Flowers axillary.

Flowers two or more together; fruit-calyces connate; seed horizontal.

Fruit-calyces two, connate towards the base, horizontally divergent; spines very short or 0; a small procumbent shrub clothed with cottony wool ... ... *biflora*

Fruit-calyces 10 to 20 together, connate into a globular woolly mass, the spines shortly protruding ... ... *paradoxa*

Flowers solitary.

Fruit-calyx densely covered with long hairs; spines divergent nearly equal; a small erect undershrub densely hairy; leaves crowded, linear, thick and soft ... ... *lanicuspis*

Fruit-calyx tomentose or nearly glabrous; spines very small; leaves linear-clavate, thick, hoary *uniflora*

Fruit-calyx tomentose, depressed; spines nearly equal, short; leaves linear, thick, mostly acute; seed horizontal ... ... *diacantha*

Fruit-calyx within dense white cottony wool, nearly globular; spines divergent, long; seed horizontal ... ... *bicornis*

III. Fruit-calyx closely sessile, much indurated with 3 to 6 spines; flowers solitary, axillary.

Fruit-calyx with 5 radiating very short spines and 5 thin appendages enveloped in dense cottony wool; leaves linear, obtuse and soft; seed horizontal (B. sclerolaenoides, F.v.M.) ... ... *Eriochiton*

Fruit-calyx with 5 to 6 unequal spines; leaves flat, cuneate-obovate, white-tomentose ... ... *Cornishiana*

Fruit-calyx with 4 to 5 long unequal spines; leaves flat, thick, linear, contracted at the base, glabrous; seed erect ... ... *quinquecuspis*

Fruit-calyx with 5 short unequal spines; leaves narrow-linear or semiterete glabrous; seed erect ... ... *echinopsila*

Fruit-calyx with 3 or 4 unequal spines; leaves semiterete glabrous; seed erect or slightly oblique ... *divaricata*

Fruit-calyx with 3, rarely 4, unequal spines, the smallest spine reduced to a tubercle; leaves linear semiterete glabrous; seed obliquely erect ... ... *bicuspis*

Babbagia.

Fruit-calyx deeply and widely excavated at the base, tubular part depressed-globular; winged appendages 2, semicircular or renate ... ... *dipterocarpa*

Fruit-calyx narrowly excavated, turgid above the tubular base; winged appendages 2, obliquely rounded or broad-cuneate, unequally developed ... ... *acroptera*

Fruit-calyx slightly excavated, tubular part cylindrical; winged appendages 5, deltoid, slightly toothed ... *pentaptera*

Salicornia.

I. Spikes usually short; calyx-lobes usually narrow. Shrubby.

Spikes thick, with acute bracts; rudimentary leaves opposite triangular; flowers three together, the middle pistillate, each of the two lateral ones with 1 stamen only. Tall, very robust ... ... *robusta*

Spikes slender, with blunt bracts; leaves inconspicuous; flowers 3 together, all bisexual. A low intricately branched shrub ... ... *arbuscula*

II. Spikes cylindrical, often elongated; calyx-lobes dilated at the end. Leaves aborted.

Flowers in fives or sevens, mostly bisexual, 2 stamens to each. Spikes elongated with blunt bracts. Herb ... ... *australis*

Flowers in threes, unisexual. A slender diffuse shrub *tenuis*

Flowers in threes, all bisexual, 1 stamen to each. A somewhat tall much branched shrub ... ... *leiostachya*

Salsola.

Rigid glabrous herb, branches spreading; flowers solitary, sessile, axillary ... ... *Kali*

Suaeda.

Glabrous herb with a woody base; flowers axillary, 2 or 3 together ... ... *maritima*

ORDER AMARANTACEAE.

    I. Leaves alternate; anthers 2-celled.

Stamens free; flowers in axillary cymes ... ... **Euxolus**
Stamens shortly united at the base.
    Flowers axillary, solitary; leaves linear, succulent ... ... **Polycnemon**
    Flowers in dense spikes or heads; leaves flat; calyx-segments wholly or in part coloured ... **Ptilotus**

    II. Leaves opposite; anthers 2-celled.

Stamens shortly united in a cup at the base; flowers in terminal elongated spikes ... ... **Achyranthes**

    III. Leaves opposite; anthers 1-celled.

Flowers in axillary clusters, stigma capitate ... **Alternanthera**
Flowers in terminal heads, stigma 2-lobed ... ... **Gomphrena**

Euxolus.

Annual, erect, rather stout, branching; leaves ovate-lanceolate, long-stalked; pericarp globular, ribbed ... ... *Mitchelli*

Polycnemon.

Stamens 5; a glabrous prostrate perennial herb, with numerous ascending branches, forming dense tufts ... ... *pentandrum*
Stamens 2, unilateral; undershrubs.
    Style very short, bifid; sepals white inside ... *diandrum*
    Style entire; sepals large, all scarious ... ... *mesembrianthemum*

Ptilotus.

    I. Calyx-segments with scarious glabrous tips, more or less covered on the outside with jointed hairs (Trichinium).

Leaves clothed with a dense stellate tomentum.
    An erect undershrub 1 to 4 ft.; spikes dense ovoid; bracts and bracteoles usually brown, obtuse, glabrous; leaves thick, obovate or oblong, on long stalks ... ... *obovatus*
    Leaves thin and less densely tomentose; bracts woolly ... ... *incanus*
Leaves glabrous, or hairy, or woolly.
    Inner segments of calyx woolly inside; segments erect, pink or red. Perennial herbs with erect stems.
        Spikes 1½ to 2 in. diameter.

Calyx-segments very rigid with short narrow tips; leaves oblong-lanceolate; stem 2 to 3 ft. ... *exaltatus*

Calyx-segments with coloured obtuse glabrous tips; leaves spathulate, all crowded at the base of the short stem *Beckeri*

Spikes 1 in. diameter; outer calyx-segments rather broad, scarcely ribbed; radical leaves oblong-spathulate, obtuse or mucronate ... *gomphrenoides*

Spikes ¾ in. diameter; stems silky-woolly, leaves linear or linear-lanceolate... *helipteroides*

Inner segments of calyx woolly inside; segments curved upwards, pink.
  Stems several decumbent or densely tufted; leaves linear, acute, glabrous; spikes solitary, nearly globular ... *erubescens*

Inner segments of calyx glabrous inside; calyx usually yellow.
  Spikes finally long and cylindrical, 1 to 2 in. diameter. Stout erect perennials, 1 to 3 ft.
    Leaves linear or lanceolate, glabrous; but the young shoots and foliage sprinkled with short hairs; bracts and bracteoles broadly ovate transparent and shining ... *alopecuroideus*

    Leaves obovate or oblong, glabrous; bracts ovate or oblong with an opaque centre ... *nobilis*

  Spikes ovoid.
    Leaves linear; spikes 2 in. diameter; bracts transparent; stems erect, 1 to 2 ft. ... *macrocephalus*

    Leaves ovate or spathulate; spikes ¾ in. diameter, bracts and bracteoles thin and shining acute; stems prostrate *spathulatus*

  Spikes globular.
    Spikes about 1 in. diameter; stamens very unequal, surrounded by a few woolly hairs. Glabrous erect; leaves linear ... *hemisteirus*

    Spikes under ½ in. diameter.
      Minute acuminate scales between the filaments; calyx red. Erect; leaves narrow-linear ... *Schwartzii*

GENERA AND SPECIES.

No scales between the filaments.
Erect glabrous herb with linear
or linear-lanceolate leaves ... *leucocoma*

An intricately slender-branching
hairy herb; leaves minute, ovate
or lanceolate ... ... *parvifolius*

Leaves small rhomboid-ovate;
bracts and bracteoles glabrous,
ovate-lanceolate; style very
short ... ... *Hoodii*

   II. Calyx segments wholly coloured and glabrous, enveloped in dense cottony wool.

Spikes cylindrical; leaves oblong; prostrate branching glabrous herb ... ... *Murrayi*

Spikes globular; leaves obovate; erect branching herb, covered with cottony wool ... ... *latifolius*

Achyranthes.

An erect spreading herb, 2 to 3 feet; leaves ovate to oblong, shortly stalked, softly pubescent; flowers green ... ... *aspera*

Alternanthera.

Glabrous herb with narrow leaves; calyx-segments and bracteoles 1½ line long, with finely pointed tips; spikes in dense clusters ... ... *triandra*

Hairy; leaves broad; calyx-segments and bracteoles shorter; spikes finally cylindrical ... ... *nana*

Gomphrena.

An erect branching annual, hoary; leaves linear; spikes globular; calyx very woolly outside; staminal tube shorter than the ovary with lobes between the filaments ... ... *Brownii*

ORDER PLUMBAGINEAE.
Plumbago.

Calyx beset with short viscid hairs. A half-climbing glabrous shrub; flowers white, sessile within 3 bracts ... ... *Zeylanica*

ORDER NYCTAGINEAE.
Boerhaavia.

Herbs; flowers in umbels; calyx viscid-hairy; stamens 1 to 4.
Prostrate; leaves orbicular-lanceolate; upper part of calyx bell-shaped, pink, about 1 line diameter ... ... *diffusa*

Ascending or half-climbing; leaves cordate-ovate, upper part
of calyx funnel-shaped, exceeding ¼ in. diameter ... *repanda*

ORDER URTICACEAE.

Flowers unisexual enclosed in a globular receptacle, closed
at the orifice by small bracts. Trees or shrubs ... **Ficus**
Flowers unisexual in axillary clusters. Herbs.
Leaves alternate, no stinging hairs. Calyx of female
flowers divided into 4 equal lobes ... ... **Parietaria**
Leaves opposite; beset with stinging hairs. Calyx of
female flowers with 2 outer segments smaller than
the two inner ... ... **Urtica**

Ficus.

Leaves on broad stalks, ovate, firm, glabrous ... ... *platypoda*
Leaves on short stalks, orbicular, firm, very scabrous above,
bordered by minute rigid teeth or callosities ... ... *orbicularis*

Parietaria.

A diffuse, pubescent, branching annual, 1 ft. or less; leaves
on slender stalks, ovate to cordate, 3-nerved from the
base; flowers few-together, in axillary cymes or almost
in clusters, within 3 or 4 bracts shortly united at the
base ... ... *debilis*

Urtica.

An erect perennial; leaves on long stalks, lanceolate somewhat cordate, acute, deeply toothed; male and female
flowers in distinct clusters... ... ... *incisa*

ORDER CASUARINEAE.

Casuarina.

1. Cone-vales prominent, keeled; fruit-wings acute.
Trees.

Branchlets ribbed, drooping, dull-green; leaf-scales 9 to 12.
Cones globular, large; valves villous inside; fruits
dark-brown ... ... *quadrivalvis*
Branchlets faintly striate, robust, erect, greyish-green;
leaf-scales 9 to 16. Cones globose, ½ in. diameter;
fruits grey ... ... *glauca*
Leaf-scales 9 or 10; cones very short, silky tomentose;
cone-valves smaller, minutely pointed; outer bark
scaly ... ... *lepidophloia*

GENERA AND SPECIES.

II. Cone-valves prominent, with a dorsal appendage. Trees.

Branchlets slender, erect, dark-green; leaf-scales 6 to 8; cones ovoid-globular, dorsal appendage thick ... ... *suberosa*

Branchlets ribbed, erect; leaf-scales 10 to 12. Cones large, ovoid-globular, dorsal appendage produced into a long rigid point ... ... *bicuspidata*

III. Cone-valves scarcely or not at all protruding; winged appendage of the seed-like fruit rounded at the end.

Branchlets nearly round, rigid; leaf-scales 4 to 5. Cones 2½ in. long, cylindrical; protuberance conical, furrowed. Tree ... ... *Decaisneana*

Branchlets ribbed; leaf-scales 4; cones oblong, protuberance of valves divided into small tubercles; fruits black. Shrub. ... ... *humilis*

Branchlets erect; leaf-scales 6 to 8. Cones ovoid-oblong, protuberance short, smooth; fruits dark-brown. Shrub *distyla*

Sub-Class II.—Choripetaleae Perigynae.

ORDER LEGUMINOSAE.

Sub-Orders.

Flowers irregular; stamens 10 or rarely less, free or united; petals imbricate.
    Upper petal or standard outside, the lowest petals united; stamens 10, united into one or two bundles, or all free; radicle curved accumbent ... ... **Papilionaceae**
    Upper petal inside; petals all disconnected; stamens 10 or fewer, all free; radicle straight **Caesalpinieae**

Flowers regular in dense globose or cylindrical spikes; petals valvate, free or united; stamens usually indefinite, free; radicle straight ... ... **Mimoseae**

Leguminosae—Papilionaceae.

I. Stamens all free; leaves simple or absent, rarely pinnate.

Standard small or narrow, ovules 4 or more ... ... **Brachysema**

Standard orbicular or renate, large; ovules 4 or more. Calyx-lobes much longer than the tube.
    Lobes imbricate; ovary sessile; pod oblong-linear; leaves simple ... ... **Isotropis**
    Lobes valvate; ovary stalked; pod globular; leaves ternate ... ... **Gompholobium**

Standard orbicular or renate, large; ovules 2.
 Pod ovoid or nearly globular; no stipules.
  Calyx-lobes longer than the tube, valvate ... **Burtonia**
  Calyx-lobes shorter than the tube, not valvate.
   Pod shortly stalked, longitudinally divided ... **Mirbelia**
   Pod stalked; leaves absent; flowers yellow ... **Sphaerolobium**
   Pod sessile, not bursting; leaves reduced to filiform phyllodia; seeds appendaged **Viminaria**
 Pod triangular, very inequilateral.
  Calyx-teeth very short; bracteoles and stipules absent; seeds strophiolate. Small shrubs ... **Daviesia**
 Pod oval-oblong, 2-valved. Small shrubs or undershrubs.
  Stipules absent; calyx more or less 2-lipped.
   Bracteoles persistent close under the calyx or adnate to it; seeds strophiolate ... **Phyllota**
   Bracteoles absent or very deciduous.
    Leaves scattered or imperfectly whorled; seeds without an appendage ... **Aotus**
    Leaves opposite, seeds with an appendage ... **Eutaxia**
    Leaves alternate or crowded; seeds with an appendage ... **Dillwynia**
  Stipules present; seeds strophiolate. Shrubs.
   Bracteoles absent or very deciduous; leaves coriaceous; pod coriaceous, turgid ... **Gastrolobium**
   Bracteoles persistent close under the calyx or adnate to it; leaves herbaceous; pod oval; petals yellow and reddish or purple ... **Pultenaea**

II. Stamens all united in a sheath, open on the upper side.

Leaves simple or absent; flowers axillary, solitary or clustered; seeds appendaged.
 Anthers of equal length; pod very flat.
  Pod winged, bursting along one edge ... **Platylobium**
  Pod not winged, bursting along both edges **Bossiaea**

Anthers 5 short and 5 long.
  Pod twice as long broad, compressed; ovules usually about 6; petals red or yellow... ... ... Templetonia
  Pod turgid; ovules usually 2; petals blue... Hovea
Leaves consisting of three leaflets; seeds appendaged. Anthers alternately long and short; pod stalked, flat.
  Flowers axillary, 2 or 3 together; ovules 6 ... Nematophyllum
  Flowers racemose, terminal; ovules 3 to 4 ... Goodia
Anthers of equal length; ovary sessile with several ovules ... ... Ptychosema
Leaves ternate or simple; flowers racemose, terminal; anthers alternately long and short; seeds without appendages; pod turgid or inflated; ovules 2 or more ... ... Crotalaria

   III. Stamens all united, more or less split into two bundles of 5 each.

Leaves pinnate; stipules present; pod jointed and separating into 1-seeded portions... ... ... Æschynomene

   IV. Stamens 9 united, the upper one free.

Leaves pinnate; stipules present.
  Pod not bursting, *muricate*, 1- or 2-seeded; seed without appendage; stipules narrow deciduous Glycyrrhiza
  Pod 2-valved; seeds usually more than 2.
    Anthers tipped with a small gland.
      Style not bearded; pod generally elongate or rarely globular, seed without appendage. Foliage with appressed forked hairs; stipules small setaceous. Herbs or shrubs ... ... Indigofera
    Anthers without glands.
      Style not bearded under the stigma.
        Pod linear, racemes terminal or leaf-opposed; seed strophiolate Tephrosia
        Pod linear, imperfectly bursting, *divided by transverse partitions*, seed without appendage; racemes axillary ... ... Sesbania
      Style bearded; pod turgid or inflated, seed without appendage. Herbs.
        Petals pointed; flowers large ... Clianthus
        Petals blunt; *pod imperfectly bursting* ... ... Swainsonia

Leaves consisting of one or three leaflets.
  Pod 1-seeded, more or less indehiscent (also Indigofera linifolia).
    Stipules linear-subulate, flowers axillary clustered ... **Lespedeza**
    Stipules large attached by a broad base; *foliage glandular-dotted.* Racemes axillary ... **Psoralea**
  Pod 2- or more-seeded.
    Flowers in axillary umbels; seeds several, without appendage.
      Lower petals blunt, leaflets 3, stipules semi-sagittate, adnate ... **Trigonella**
      Lower petals pointed, leaflets 5, the two lowest taking the place of stipules ... **Lotus**
    Flowers in axillary racemes, or the peduncles 1- to 3-flowered.
      Seeds strophiolate separated by thin septa ... **Kennedya**
      Seeds without an appendage (also Indigofera monophylla and Swainsona unifoliolata).
        Flowers racemose, pod septate between the seeds. Climbing or twining herbs ... **Glycine**
          Trees with conical prickles ... **Erythrina**
          Pod without septa, a training or twining herb ... **Rhynchosia**
        Flowers in clusters, or 2 or 3 at the end of the peduncle. Twining herbs.
          Style beardless ... **Galactia**
          Style bearded under the stigma **Vigna**

Leguminosae—Caesalpinieae.

Leaves simply pinnate or reduced to phyllodia; Stamens 10, 7 or all perfect; anthers opening in terminal pores or short slits. Tall shrubs; flowers yellow ... **Cassia**

Leaves unequally pinnate; perfect stamens 3; anthers opening by longitudinal slits; staminodia 2, small; style large and petal-like ... **Petalostylis**

Leaves of 2 leaflets or 2-lobed. Trees ... **Bauhinia**

Leguminosae—Mimoseae.

Stamens 5, anthers without any gland, petals valvate **Neptunia**
Stamens indefinite. Trees or shrubs ... **Acacia**

Brachysema.

A leafless, silky-pubescent, often spinescent shrub, flowers crowded on short radical stalks ... ... *Chambersi*

Isotropis.

Leaves of 1 leaflet, articulate on the petiole; calyx tomentose.
   Leaflets ovate or oblong, very obtuse ... ... *atropurpurea*
   Leaflets terete, channelled above ... ... *Wheeleri*
   Leaflets lanceolate-oblong, on long stalks ... *Winneckei*

Gompholobium.

Depressed glabrous herb; leaflets from broad- to narrow-linear; flowers large, red or yellow, generally 2 or 3 together, on shorter stalklets; calyx and lower petals glabrous ... ... *minus*

Burtonia.

Erect, very hirsute shrub; leaves pinnate; racemes terminal ... ... *polyzyga*

Mirbelia.

Rigid leafless shrub; branchlets thorny ... ... *oxyclada*

Sphaerolobium.

Undershrub; stems terete; flowers in irregular racemes ... ... *vimineum*

Viminaria.

Tall glabrous shrub; flowers in long terminal racemes *denudata*

Daviesia.

Flowers 2 or 3, umbellate; leaves linear-lanceolate ... *arthropoda*
Flowers racemose axillary; branches slightly angular; bracts small.
   Flowering branches with narrow-elongate, rigid, leaves. A tall glabrous shrub ... ... *corymbosa*
   Flowering branches spinescent and leafless ... *horrida*
Flowers in axillary clusters, few together or sometimes only 1.
   Leaves flat, rigid, lanceolate, vertical, pungent, sessile with a broad base; bracts large ovate. *pectinata*
   Leaves flat, ovate-linear, horizontal, simply sessile, pungent; branches spinescent; bracts very small ... ... *ulicina*

Leaves cylindrical, pungent; bracts very small.
Leaves articulated on the branchlets.
    Lower petals not much curved, obtuse   *genistifolia*
    Lower petals much curved, acute; leaves ½ in. long, dilated upwards ... ... *incrassata*
Leaves continuous with the branchlets, few, prickle-like, ¼ in. long; lower petals much curved, acute... ... ... *brevifolia*

Aotus.

Heath-like shrub; leaves linear; flowers yellow with a purple keel; calyx pubescent ... ... *villosa*

Phyllota.

Flowers singly sessile within clusters of terminal leaves. A small shrub, with small linear pointed leaves; bracteoles ovate, shorter than the hairy calyx ... ... *pleurandroides*

Flowers crowded in short leafy spikes at end of branchlets ... ... *Sturtii*

Eutaxia.

A low glabrous intricate shrub; ovary stalked, style subulate, pod turgid; leaves small oval to linear   *empetrifolia*

Dillwynia.

I. Calyx gradually attenuated at the base; petals deciduous; standard on a long claw, the lamina fully twice as broad as long.

Lower petals pointed, nearly as long as the lateral ones. A small shrub with short spreading hairs; leaves short, spreading, linear-cylindrical; flowers orange coloured and reddish, in short racemes on long peduncles... ... ... *hispida*

Lower petals blunt, much shorter than the lateral ones,
    Flowers mostly terminal in sessile corymbs or clusters. An erect heath-like shrub ... ... *ericifolia*
    Flowers mostly axillary, solitary or in few-flowered clusters on short racemes. An erect undershrub; leaves small linear-cylindrical, furrowed above ... ... *floribunda*

II. Calyx blunt at the base; petals persistent; claw of the standard shorter than the calyx, lamina rather broader than long.

Flowers several in terminal corymbs; leaves under ½ in. long, rather slender; upper lip of calyx emarginate ... ... *cinerascens*

Flowers few in terminal corymbs, longer than the upper leaves; leaves under ¼ in. long, rather thick and obtuse; upper lip of calyx shortly 2-lobed ... ... *patula*

Gastrolobium.

Flowers axillary, solitary; leaves ovate; calyx ¼ in. long ... ... *elachistum*

Flowers racemose, large; leaves ovate; calyx nearly ½ in. ... ... *grandiflorum*

Pultenaea.

1. Flowers in terminal sessile heads.

Bracteoles adnate to the calyx-tube; *leaves flat*.
  Heads surrounded by imbricate bracts, the inner ones longer than the flower-stalks.
    Heads rather large; leaves cuneate-oblong, minutely pointed, about 1 in. long; calyx silky hairy; stipules minute; bracteoles small linear. Rather tall robust ... ... *daphnoides*

    Heads small; leaves under ½ in. long; pod hairy. A small erect shrub ... ... *stricta*

    Heads dense; *stipules large*, brown, pointed; pod very acute. Softly villous, rather low shrub ... ... *mucronata*

    Heads surrounded by bracts, shorter than the flower-stalks; leaves obovate emarginate, scabrous above; stipules spreading ... ... *scabra*

Bracteoles adnate to the calyx-tube; *leaves linear-cylindrical*. Erect, rather tall, softly hairy; bracteoles narrow, keeled ... ... *mollis*

Bracteoles free from the base of the calyx.
  Leaves rigid, *pungent*; stipules conspicuous.
    Leaves lanceolate; flowers stalked ... ... *rigida*
    Leaves trigonous; flowers sessile ... ... *acerosa*
    Leaves narrow-linear, mucronate; *stipules imbricate* ... ... *vestita*
    Leaves linear-terete, blunt, channelled; bracteoles very narrow, long; stipules narrow pubescent; heads short, leafy ... *canaliculata*

11. Flowers in terminal clusters, lengthening into leafy spikes; calyx silky-pubescent; stipules small.

Flowers sessile or nearly so; leaves broadly obovate to linear-cuneate; bracteoles near the top of the calyx-tube ... *largiflorens*

Flowers distinctly stalked; leaves small, linear, channelled; bracteoles at base of calyx, linear-lanceolate ... *laxiflora*

111. Flowers solitary, terminal, surrounded by imbricate bracts.

Bracteoles adnate to calyx. Silky-pubescent; leaves linear-terete ... *prostrata*

Bracteoles free. Villous; leaves linear or lanceolate, acute ... *involucrata*

IV. Flowers axillary.

Bracteoles adnate to the calyx-tube.
  Flowers solitary, scattered.
    Flowers on long stalks; prostrate; leaves linear to oblong-lanceolate; calyx-lobes acute; bracteoles linear ... *pedunculata*
  Flowers crowded into short leafy-spikes or racemes at or near the end of the branchlets.
    Leaves elliptical-oblong or linear, about ½ in. long; bracteoles linear-subulate, ciliate. Dwarf, softly hairy ... *humilis*
    Leaves oblong-linear, 2 lines long; bracteoles narrow-lanceolate, viscid when young. Rather tall, with lax, somewhat drooping branchlets, softly hairy ... *graveolens*

Bracteoles free from the base of the calyx.
  Flowers solitary, scattered; sessile. Dwarf, prostrate, softly pubescent; leaves narrow-linear or terete ... *tenuifolia*
  Flowers forming short leafy spikes or racemes.
    Branchlets pubescent or villous.
      Leaves small, broad, rigid; flowers crowded into almost leafy clusters; calyx-lobes broad, pointed. A diffuse shrub ... *densifolia*
      Leaves acute, hirsute or ciliate with long hairs; calyx-lobes acuminate. Rather tall, erect ... *villifera*
    Branchlets shortly hairy, viscid; leaves linear-terete, slender; calyx-lobes acuminate. Rather tall, erect ... *viscidula*

Platylobium.

Pods sessile, pedicels concealed by bracts; leaves deltoid to cordate-ovate, pointed, opposite; flowers yellow and red, solitary. A dwarf, straggling erect shrub ... ... *obtusangulum*

Pods stalked, pedicels longer than the bracts; leaves deltoid ... ... *triangulare*

Bossiaea.

I. Leaves alternate; branchlets cylindrical or angular.

Prostrate. Leaves ovate or oblong; flowers yellow on long stalks ... ... *prostrata*

Erect. Leaves cordate-lanceolate to linear, pungent-pointed; upper calyx-lobes much longer than the lower ... ... *cinerea*

II. Branchlets leafless, much compressed; glabrous, slightly indented at the nodes.

Lower petals almost as long as the standard ... ... *riparia*

Lower petals longer than the standard; flowers large, pink ... ... *Walkeri*

III. Branchlets leafless, cylindrical, glabrous, furrowed.

Small intricately branched shrub; calyx-lobes ciliated; flowers small, reddish, solitary or two together ... ... *Battii*

Templetonia.

I. Branchlets leafy; stipules minute or inconspicuous.

Leaves cuneate or obovate, obtuse or emarginate. Tall shrub with large, red, rarely yellow, flowers *retusa*

Leaves narrow-linear, flat or slightly channelled. Dwarf shrub, with small yellow flowers; pod stalked ... ... *Muelleri*

II. Branchlets leafless.

Stipules prickly, recurved; branchlets cylindrical. Low rigid shrub; pod on a stalk longer than calyx *aculeata*

Stipules minute or absent; flowers very small, yellow. Branchlets cylindrical, furrowed, glabrous; flowers aggregated in spike-like racemes. Tall shrub ... ... *egena*

Branchlets broadly flattened, furrowed, glabrous, much spreading; flowers scattered. Tall shrub *sulcata*

E

Hovea.

Tall erect shrub; leaves obtuse at both ends; pod hairy ... *longifolia*

Prostrate or slightly ascending, subshrubby; lower leaves ovate, upper lanceolate; pod nearly glabrous ... *heterophylla*

Nematophyllum.

A tall slender shrub; leaflets long linear-terete ... *Hookeri*

Goodia.

Leaflets usually cuneate-ovate; pod about thrice as long as broad, gradually narrowed at the base, pod-valves reticulated. Tall shrub, flowers yellow ... *lotifolia*

Leaflets usually obcordate-ovate; pod shorter and smoother, suddenly narrowed at the base ... *medicaginea*

Ptychosema.

Small weak herb; leaves minute, penninerved; racemes remotely few-flowered; petals pinkish-white ... *anomalum*

Procumbent; leaflets 3, minute, not veined; peduncles 1-flowered; keel dark purple, other petals yellow ... *trifoliolatum*

Crotolaria.

Leaves simple, continuous with the short stalk.
    Leaves oblong, the upper ones linear; petals short *linifolia*
    Leaves obovate, retuse; petals and pod much longer than the calyx; flowers in dense racemes *Mitchelli*

Leaves simple, the stalk articulate above the middle, ovate. Softly tomentose shrub; flowers very large, yellowish-green ... *Cunninghami*

Leaflets 3 or 1, narrow, ovules many, pod oblong longer than the calyx; flowers small, remote, in few-flowered racemes. A slender, erect, softly tomentose herb ... *dissitiflora*

Leaflets three, oblong-cuneate; ovules 2; pod small, orbicular ... *medicaginea*

Æschynomene.

Erect annual, 1 to 2 feet; leaflets 40 to 60 ... *Indica*

Glycyrrhiza.

Erect herb, or undershrub; leaflets 9 to 11; flowers small, in racemes ... *psoraleoides*

Indigofera.

I. Leaves simple or of one leaflet.

Leaves simple, nearly sessile, linear; flowers in short spikes, calyx-lobes longer than tube; pod globose, 1-seeded. Herb ... ... *linifolia*

Leaflet obovate, articulate on the stalk; calyx-teeth all short ... ... *monophylla*

II. Leaves consisting of several pairs of leaflets.

Calyx-lobes very much longer than the tube. Herbs.
  Pod short; seeds 2; flowers very small, in short dense spikes ... ... *enneaphylla*
  Pod linear, slender; ovules and seeds several.
    Calyx much shorter than corolla; pod viscid *viscosa*
    Calyx as long as the corolla; plant with spreading hairs ... ... *hirsuta*

Calyx-lobes very short. Slender shrubs.
  Calyx-lobes inconspicuous; pod glabrous; leaflets 9 to 17, orbicular-ovate to linear-elliptical; flowers red in racemes about as long as leaves *australis*

Calyx distinctly but shortly toothed.
  Hoary shrub; pod pubescent, leaflets shortly stalked ... ... *brevidens*
  Tomentose; pod glabrous; leaflets sessile ... *coronillifolia*

Tephrosia.

Leaflets 7 to 11; seeds transversely oblong ... ... *purpurea*
Leaflets 4 to 8; seeds spherical ... ... *sphaerospora*

Sesbania.

A very tall herb; leaflets in 20 to 50 pairs, the stalk up to 1 foot long sometimes beset with small prickles; calyx-teeth very short ... ... *aculeata*

Clianthus.

A wide spreading, softly hairy herb; leaflets 15 to 21; standard 2½ in. long, red, rarely white, with a black shining blotch at the base ... .. *Dampieri*

Swainsonia.

I. Standard with two prominent callosities near the base.

Pod with a stalk-like base. Subshrubby.
  Calyx tomentose; flowers pink, large, in long racemes. Pod large, inflated, valves thin. Tall, somewhat hairy ... ... *Greyana*

Calyx glabrous, flowers pink or white, stipules
small ... ... *coronillifolia*
Calyx glabrous, flowers violet; stipules large ... *colutoides*
Pod sessile.
  Keel simply curved; style slender; ovary silky.
    Calyx-teeth lanceolate; leaflets more than 9.
      Dwarf, somewhat hoary; leaflets linear
      or oblong; flowers large, brick-red,
      few on long stalks; stipules lanceolate *phacoides*
      Stems rigid; villous; leaves obovate;
      stipules broad ... ... *Burkitti*
    Calyx-teeth subulate or very short; leaflets
    usually less than 9; flowers few in short
    racemes.
      Slightly hoary; leaflets 5 to 9, obovate
      to cuneate-oblong ... ... *oligophylla*
      Softly villous; *raceme dense, ovoid be-
      fore expansion* ... ... *Burkei*
      Glabrous; leaflets 3 to 5, lanceolate ... *oroboides*
  Keel twisted; style thick; ovary glabrous;
  leaflets about 5 ... ... *campylantha*

II. Standard without callosities.

Keel twisted; leaflets 11 to 21 or more, lanceolate;
petals large, violet; pod sessile, oblong; stipules
rather large. Stems ascending or prostrate;
more or less glabrous ... ... *procumbens*
Keel simply curved; style hooked, bearded near
the stigma.
  Leaflets 7 to 11, narrow; petals orange; stipules
  large ... ... *stipularis*
  Leaflets very small, oblong; ovary silky-hairy;
  petals blue... ... ... *Oliverii*
Keel simply curved. Style not hooked, bearded on
inside.
  Leaflets 9 to 15, linear- to ovate-elliptical,
  stipules rather broad; flowers violet in short
  racemes or umbels on long peduncles; calyx
  with appressed black hairs. Prostrate, somewhat hairy... ... ... *lessertifolia*
  Leaves reduced to 1 obovate leaflet; stipules
  minute, deltoid ... ... *unifoliolata*
Keel simply curved. Style not hooked, with a tuft
of hairs, besides the beard under the stigma.
  Leaflets very small, numerous, from obcordate
  to cuneate-ovate; stipules minute; flowers
  violet in long racemes; pod ovate-globular,
  sessile ... ... *microphylla*

GENERA AND SPECIES.

Leaflets numerous, stalked, ovate to elliptical; stipules broad, obtuse; flowers yellow; pod acute with a stalk-like base ... ... *laxa*

Lespedeza.
Shrubby, densely velvety-tomentose or woolly ... *lanata*

Psoralea.
I. Leaflets three, entire.
Erect glabrous or slightly hairy; leaflets large, 1 to 3 in. long, lanceolate or oblong-elliptical; racemes spike-like on very long stalks; petals pink; bracts ovate-lanceolate; pod blackish, somewhat rough ... ... *adscendens*

Dwarf diffuse, flowers and leaves smaller; bracts orbicular-cordate; pod beset with soft hairs ... *parva*

II. Leaflets three, toothed (also *P. leucantha*).
Erect lax, softly hairy; calyx short, its lobes equal; leaflets lanceolate- to rhomboid-ovate; racemes on long stalks; petals pink ... ... *patens*

Prostrate, white-tomentose; calyx elongated, its lateral lobes short; flowers blue in spike-like racemes ... ... *eriantha*

III. Leaf of one leaflet. Tall undershrubs.
Pubescent or villous; flowers in subglobular racemes *balsamica*

Glabrous or slightly hoary; flowers in loose elongated racemes; leaflets oblong or lanceolate; flowers white ... ... *leucantha*

Trigonella.
Prostrate annual, sweet-scented; flowers pale yellow in axillary sessile clusters; pods narrow, compressed ... ... *suavissima*

Lotus.
Prostrate or decumbent; petals yellow; calyx-lobes about as long as the tube; flowers several in the umbel ... ... *corniculatus*

Tall herb; petals white or pink; calyx-lobes usually longer than the tube; flowers several in each umbel; leaflets broadly linear to obovate, ½ in. or more long ... ... *australis*

Decumbent; leaves small and broad; flowers solitary or 2 or 3 together, small, dark-reddish *var. Behrianus*

Kennedya.

Erect shrub with elongate twining branchlets; leaflets solitary, large, ovate or ovate-lanceolate; flowers blue, rarely white, numerous, racemose or paniculate ... *monophylla*

Prostrate herbs with leaves of three leaflets.
    Flowers rather large, crimson, 1 or 2 on each peduncle; leaflets orbicular to ovate, with undulate margins ... *prostrata*
    Flowers racemose, violet ... *prorepens*

Glycine.

I. Lateral leaflets close to the terminal one.

Stem and branches slender, twining, hirsute with reflexed hairs; leaflets of the upper leaves narrow; upper calyx-lobes deeply cleft; pod narrow, compressed; flowers in racemes in the upper axils, or clustered or imperfect and smaller in the lower axils ... *clandestina*

Stems and branches short and often prostrate; leaflets of the upper leaves broad; upper calyx-lobes much united ... *Latrobeana*

Stems short decumbent; leaflets lanceolate or oblong on a long hairy petiole; pod falcate, very hairy ... *falcata*

II. Lateral leaflets distant from the terminal one.

Leaflets of the lower leaves short and broad; stems and branches elongate beset with reflexed hairs.. *tabacina*

Leaflets linear-acute; stems beset with appressed hairs; pod densely silky-pubescent ... *sericea*

Leaflets ovate or oblong, obtuse; stems prostrate or twining, softly tomentose or villous ... *tomentosa*

Erythrina.

Leaflets 3, broadly obcuneate, 2- or 3-lobed, 3 or 4 in. broad ... *vespertilio*

Rhynchosia.

Leaflets broadly ovate-rhomboid, about 1 in. long; pod falcate, longer than broad, hairy ... *minima*

Galactia.

Glabrous or with spreading hairs; flowers few, small *tenuiflora*

Vigna.

Glabrous or slightly pubescent; leaflets lanceolate or linear; flowers pale yellow ... *lanceolata*

Cassia.

I. Stamens 7, perfect, of which 2 or 3 lower ones are larger or on longer filaments; staminoidia 3, small.

Raceme short, almost corymbose; pod thick; seeds horizontal ... ... *Sopherae*
Raceme elongated on long axillary peduncles; pods very flat; bracts large deciduous.
   Pubescent; leaflets 9 to 15 pairs.
      Stipules ovate, cordate, rigid; bracts broad, obtuse... ... ... *venusta*
      Stipules narrow, bracts acuminate ... ... *notabilis*
   Glabrous; leaflets 4 to 5 pairs, oblong-linear; stipules small subulate; bracts broad, obtuse *pleurocarpa*

II. Stamens 10, all perfect.

Leaflets flat.
   Very glutinous, otherwise glabrous, leaflets 8 to 10 ... ... *glutinosa*
   Glaucous; stipules leafy half-cordate ... ... *pruinosa*
   Hoary, becoming glabrous with age; leaflets rarely more than 3, ovate or ovate-oblong; pod $\frac{1}{2}$ in. broad, stalked... ... ... *desolata*
   Glabrous; leaflets 6 to 10, dull green on both sides, lanceolate to linear-elliptical, somewhat concave ... ... *Sturtii*
Leaflets cylindrical or linear-terete, more or less channelled.
   Leaf-stalk quite narrow; leaflets 6 to 12, hoary-white ... ... *artemisiodes*
   Leaf-stalk dilated and vertically flattened; leaflets 1 or 2 pairs, green or almost glabrous ... *eremophila*
Leaflets usually undeveloped, reduced to vertically flattened phyllodia.
   Phyllodia slender and green; peduncles 1- or 2-flowered; pod very much curved ... ... *circinata*
   Phyllodia thick hoary; peduncles several-flowered; pod flat, broad, straight ... ... *phyllodinea*

Petalostylis.

Erect glabrous shrub, leaflets 11 to 30; stalks 1-flowered; petals spreading, nearly equal, large, obovate, orange ... ... *labicheoides*

Bauhinia.

Calyx-tube very short, the free part deeply lobed ... *Leichhardtii*
Calyx-tube turbinate, the free part shortly lobed ... *Carronii*

Neptunia.

Erect herb, leaves bipinnate; pod orbicular, 1-seeded   *monosperma*
Peduncles *long, slender*; pod oblong, several-seeded...   *gracilis*

Acacia.

Key to the chief groups.

Leaves reduced to phyllodia or wanting (except in the seedling plant which has pinnate leaves).
  Flowers in globular heads.
   I. Phyllodia absent, branchlets spinescent.
   II. Phyllodia spinescent, cylindrical or lanceolate, scattered.
   III. Phyllodia whorled, spinescent.
   IV. Phyllodia more or less cylindrical, elongate, not pungent.
   V. Phyllodia more or less ovate, if linear very short, more or less oblique, usually acutely pointed, and 1-nerved.
   VI. Phyllodia flat, prominently 1-nerved, long and narrow.
   VII. Phyllodia flat, 2-nerved.
   VIII. Phyllodia flat, 3- to 5-nerved.
   IX. Phyllodia flat, with numerous striae.
  Flowers in cylindrical spikes.
   X. Phyllodia rigid, spinescent.
   XI. Phyllodia not spinescent.
Leaves all bipinnate.
The seeds are strophiolate and longitudinally placed in the pods except when otherwise stated.

List of species having:—

Large bracts to the unexpanded heads:—*iteaphylla, spinescens, suaveolens, sublanata.*
Spinescent stipules:—*armata, aspera, Farnesiana, oxycedrus, pyrifolia, Sentis, strongylophylla.*
Spinescent branchlets:—*acanthoclada, continua, erinacea, Peuce, spinescens.*
Viscid branchlets:—*dodonaeifolia, montana, verniciflua.*
Transverse or oblique seeds:—*anceps, aneura, Burkittii, craspedocarpa, cyperophylla, dictyophleba, impressa, Kempeana, lysiphloia, minutifolia, Murrayana, notabilis, retivenea, Sentis, Spilleriana, spondylophylla, stipuligera, strongylophylla, suaveolens.*
Funicle doubly bent around the seed:—*cyclopis, melanoxylon, notabilis, retinodes, Wattsiana.*
No strophiole:—*estrophiolata, romeriformis.*

I. Phyllodia absent, the branchlets resembling spinescent phyllodia.

Branchlets shortly decurrent with stem, terete, 1 to 2 in. long; flowers rather large, many in each solitary head; pod narrow, twisted, constricted between the seeds ...   .   *continua*

Pod very flat and broad; branchlets crowded, slender and rigid, 2 to 4 in. long   ...   ...   *Peuce*

Branchlets articulate on the stem; flowers very small and few in each solitary head; pod narrow, twisted, constricted between the seeds; flower buds enclosed in prominent bracts ... ... *spinescens*

 II. Phyllodia cylindrical or linear, spinescent, scattered (also *A. lanigera*).

Phyllodia faintly many-veined, cylindrical-linear; flower-heads 1 to 4 on short stalks; sepals 5, spathulate; pod narrow curved ... ... *colletioides*

Phyllodia prominently 1-veined; flower-heads 1 or 2 on stalks as long as phyllodes; sepals 5, spathulate; pod narrow, twisted, contracted between the seeds ... ... *genistioides*

Phyllodia prominently 1-veined, linear or narrow-lanceolate, broad at the base; flower-heads solitary on stalks as long as phyllodes; calyx shortly 4-lobed; pod linear, straight or curved *rupicola*

 III. Phyllodia terete-cylindrical or short and compressed, sharp-pointed, clustered or whorled. (Also *A. verticillata*). Sepals and petals 5, united or free.

Phyllodia prismatically cylindrical, 3 to 6 in a cluster; heads 1 to 3 on long stalks; sepals free; pod curved, narrow constricted ... ... *tetragonophylla*

Phyllodia linear-terete, hairy, 9 to 13 in a whorl; heads 1 or 2 on stalks longer than the phyllodes; calyx 5-lobed; pod, short, broad, curved, hairy; seeds oblique, funicle short, straight ... ... *spondylophylla*

Phyllodia subulate, sulcate, recurved-pointed, 8 to 10 in a whorl... ... ... *lycopodifolia*

Phyllodia minute, rhomboid-ovate, compressed, spinulate-point lateral; heads 1 on long stalks; corolla deeply cleft in lanceolate segments; pod lanceolate-oblong; seeds oblique; funicle long, straight ... ... *minutifolia*

 IV. Phyllodia narrow-linear or subulate, terete or subangulated with short innocuous recurved points. Heads 1 or 2 together, rarely in a few-flowered raceme; petals and sepals 5; stipules rarely present; pod elongate, more or less curved, or twisted and constricted between the seeds.

Phyllodia with 2 or more prominent nerves.
 Phyllodia compressed-filiform, 1 vein on each side; sepals ciliate on the margin; heads 3 or 4 in a short raceme ... ... *calamifolia*

Phyllodia linear-subulate, slightly flattened, sparingly and shortly hairy, 1 vein on each side; heads solitary on short stalks; corolla twice as long as calyx; pod straight ... ... *scirpifolia*

Phyllodia linear-subulate, flat, 3 to 5 veins on each side, calyx-lobes spathulate, hair-tufted; heads 2 to 4 on very short stalks ... ... *rigens*

Phyllodia of the branchlets linear, acutely tetragonal, 4-veined; the older phyllodia varying from oblong to linear-spathulate, prominently 1-veined; heads 2 together; sepals free, spathulate, with ciliate margins; pod linear straight ... ... *gonophylla*

Phyllodia longitudinally striated; linear-subulate.

Phyllodia, slightly prismatic; *heads* 2, *sessile*; ovary hairy; pod broad-linear, twisted ... *sessiliceps*

Phyllodia slightly flattened; heads solitary; pod flat, broad, curved, *valves almost membranous* ... ... *papyrocarpa*

Phyllodia thickly filiform, acute; heads 2 to 4... *Gilesiana*

v. Phyllodia flat, small, and usually broad, more or less oblique, usually acutely pointed and 1-nerved. Petals and sepals 5; peduncles 1-headed, one or two together.

Stipules spinescent; phyllodia 1-nerved.

Phyllodia semi-ovate or-lanceolate, *undulate*; pod linear, straight or curved, hairy; calyx-lobed. Tall shrub. ... ... *armata*

Phyllodia cordate-orbicular; pod flat, narrow-oblong; sepals free; *seeds transverse* ... ... *strongylophylla*

Phyllodia lanceolate-oblong or linear, *penni-reined*; pod flat, broad-oblong; sepals free; *seeds transverse*. Tree ... ... *Sentis*

Phyllodia oblong-linear; pod linear, curved, *glandular hispid*; calyx deeply cleft ... ... *aspera*

Stipules setaceous; phyllodia 1-nerved.

Phyllodia obovate or cuneate-oblong; *branchlets spinescent*; pod narrow, spirally coiled; *strophiole cup-shaped* ... ... *acanthoclada*

Phyllodia semilanceolate; pod linear, flat, contracted between the seeds; calyx cleft; *strophiole absent* ... ... *vomeriformis*

Stipules minute or wanting (also *A. Sentis* partly).

Phyllodia obovate-oblong; pod oblong; seeds *transverse*; calyx toothed; *branchlets spinescent* ... ... *erinacea*

Phyllodia ovate or rhomboid-orbicular; pod linear, twisted; sepals free ... ... *obliqua*
Phyllodia linear, obliquely spathulate; pod linear, curved or twisted; sepals free... ... *lineata*
Phyllodia broad-triangular, 3- to 5-*nerved*; pod linear, twisted; calyx short-toothed; *branchlets woolly*; bracts acuminate-setaceous longer than the young buds ... ... *sublanata*
    Pubescent; bracts very obtuse, short ... *pravifolia*
Phyllodia oblong; pod linear, twisted; sepals linear-spathulate, ciliate; petals acute ... *acinacea*
Phyllodia broadly ovate or oblong, rather large, *sessile by a broad base, and decurrent on the stem*, glaucous; pod broad-oblong, stalked, seeds transverse; calyx toothed ... ... *anceps*

  VI. Phyllodia flat, prominenently 1-nerved, usually long and narrow; sepals and petals usually 5, united or free; stipules usually absent (*A. Sentis* partly).

Flower-heads, one or 2 together, on stalks shorter or longer than phyllodia.
  Branchlets and young foliage viscid (also *A. montana* and *A. verniciflua*).
    Phyllodia oblong-linear or lanceolate, calyx-lobed; corolla smooth, deeply cleft; pod elongate, flat, straight; funicle with 2 or 3 short close folds under the seed... ... *dodonaeifolia*
  Branchlets and young foliage with adpressed hairs or glabrous.
    Peduncles shorter than the phyllodia (also *A. salicina*).
      Phyllodia broad- or linear-lanceolate; petals and sepals free; funicle straight; pod very narrow, curved ... *microcarpa*
      Phyllodia ovate or ovate-oblong; petals and sepals free; funicle with 1 fold below the seed; pod broadly linear or narrow-elliptical ... ... *brachybotrya*
    Peduncles longer than the phyllodia, seeds oblique; otherwise like *A. brachybotrya*... *Spilleriana*
Flower-heads racemose; unexpanded heads enclosed in large concave bracts; petals free, smooth; funicle very short, hardly folded.
  Phyllodia lanceolate; sepals linear-spathulate; seeds transverse ... ... *suaveolens*
  Phyllodia broadly linear, pointed; sepals setaceous; seeds longitudinal ... ... *iteaphylla*

Flower-heads racemose; without conspicuous bracts; seeds transverse.
    Phyllodia linear; funicle shortly folded under the seed .... *Murrayana*
    Phyllodia narrow- or broad-elliptical; funicle in a double fold around the seed ... ... *notabilis*

Flower-heads racemose; without conspicuous bracts; seeds longitudinal.
    Funicle in a double fold around the seed; corolla-lobes acute.
        Phyllodia linear-lanceolate, acute; calyx-lobes short, broad, ciliate; pod elongate, broad-linear, straight ... ... *retinodes*
        Phyllodia oval; calyx toothed, ciliate; pod narrow-oblong, flat ... ... *Wattsiana*
    Funicle straight, or more or less folded under the seed.
        Phyllodia elliptic-lanceolate, 3 to 6 inches long; petals glabrous; calyx lobed, ciliate; funicle short straight ... ... *pycynantha*
        Phyllodia linear-spathulate, 2 to 5 in., corolla-lobes acute, hairy; funicle short, straight; calyx 5-cleft ... ... *hakeoides*
        Phyllodia oblong-linear or-lanceolate, 2 to 5 in.; calyx truncate or 5-toothed; funicle with several close and short folds ... *salicina*
        Phyllodia ovate or ovate-oblong; petals smooth; sepals linear-spathulate; funicle once-folded ... ... *brachybotrya*
        Phyllodia ovate, *pungent-pointed; stipules thorny* ... ... *pyrifolia*
        Phyllodia lanceolate, 1 to 2 in., with a thick margin; petals 4, deeply cleft, smooth; funicle very short, scarcely folded; pod broad-linear, compressed, with a thick margin; heads of 2 to 6 pale-yellow flowers ... ... *myrtifolia*

  VII. Phyllodia flat, 2 veined; glandular-dotted. Tall viscid shrubs; flower-heads 2 together, stalked; corolla and calyx 5-lobed; pod linear, straight; seeds longitudinal; funicle shortly folded.

Phyllodia oblong-lanceolate; pod viscid ... ... *verniciflua*
Phyllodia linear-lanceolate; calyx and pod hairy ... *montana*

GENERA AND SPECIES.

VIII. Phyllodia flat, 3 to 5 prominently nerved; sepals and petals 5, free or united (also *A. sublanata*).

Flower-heads 1 or 2 together; funicle more or less folded under the seed.
    Phyllodia broad- to narrow-elliptical; corolla smooth, lobed; calyx deeply toothed, ciliate; seeds transverse ... ... *impressa*
    Phyllodia linear; sepals free; seeds without strophiole ... ... *estrophiolata*
    Phyllodia narrow-elliptical to ovate, glutinous; seeds tranverse ... ... *craspedocarpa*
    Phyllodia linear-lanceolate, pungent; petals smooth; sepals narrow spathulate; pod flat, margins thick ... ... *cochlearis*
    Phyllodia with 3 prominent nerves and conspicuous reticulations; seeds transverse.
        Phyllodia elliptic-lanceolate; corolla smooth, 5-lobed ... ... *dictyophleba*
        Phyllodia oval-oblong; petals free, hairy at the tips ... ... *retivenea*

Flower-heads in racemes.
    Phyllodia narrow-linear to linear-oblong, sepals and petals free, smooth; funicle folded under the seed ... ... *trineura*
    Phyllodia broadly elliptical to lanceolate; petals united; calyx-lobes ciliated; funicle doubly folded around the seed ... ... *cyclopis*
    Phyllodia oblong or broad-lanceolate, falcate; calyx 5-toothed; funicle doubly folded around the seed ... ... *melanoxylon*

IX. Phyllodia flat, many-streaked; sepals and petals 5, free or united.

Flower-heads in racemes; corolla 5-lobed; funicle shortly folded under the seed. Trees.
    Phyllodia broad- or linear-lanceolate, recurved-pointed; calyx 5-cleft, ciliate ... ... *homalophylla*
    Phyllodia linear, 6 inches or more long; calyx and corolla lobed hairy ... ... *stenophylla*

Flower-heads 1- to 4-clustered.
    Funicle short, not twisted; calyx, corolla, and ovary hairy.
        Phyllodia linear, recurved-pointed; heads nearly sessile, calyx with deeply cut spathulate segments; pod coiled or twisted; strophiole orange-coloured, enveloping half the seed ... ... *Osswaldi*

Phyllodia linear, 6 in. long; calyx lobed; pod twisted ... *coriacea*

Funicle twice or thrice folded beneath the seed.
  Phyllodia oblong or linear, narrowed at the base; peduncles short, glabrous; petals thin; sepals narrow-linear spathulate; pod linear, longitudinally streaked ... *sclerophylla*

  Phyllodia linear-cuneate; peduncles very short, mealy-tomentose; petals acute, hairy; sepals linear-spathulate, hairy; pod linear, contracted between the seeds *farinosa*

  Phyllodia oblong-linear; petals lobed, smooth; calyx thin, shortly lobed ... *Whanii*

  Phyllodia narrow-lanceolate, sharp-pointed; pod narrow, flexuous, hairy ... *lanigera*

x. Flowers in cylindrical spikes; phyllodia rigid, 2- to 4-nerved, spinescent; calyx and corolla 4-lobed.

Phyllodia whorled, acicular, linear; pod flat, broadly linear; stipules minute ... *verticillata*

Phyllodia scattered, narrow-lanceolate from a broad base; pod flat, broadly linear; stipules short, pungent ... *oxycedrus*

Phyllodia scattered, linear-lanceolate; flower-heads small and nearly sessile ... *rhigiophylla*

xi. Flowers in cylindrical spikes; phyllodia not spinescent; petals and sepals 5, rarely 4, free or united; funicle with 1 or few short folds; spikes solitary or in pairs.

Stipules conspicuous; phyllodia 3- to 5-nerved; seeds obliquely transverse.
  Leaves ovate-elliptical, recurved-pointed; calyx, corolla and ovary hairy; calyx lobed; seeds transverse ... *stipuligera*

  Leaves linear-oblong, obliquely pointed; calyx, corolla and ovary smooth; calyx cleft to near the base; seeds oblique ... *lysiphloia*

Stipules absent or inconspicuous.
  Phyllodia with 3 to 5 prominent nerves; calyx short, toothed.
    Phyllodia elliptic-lanceolate, straight; seeds longitudinal; calyx 4-lobed ... *longifolia*
    Phyllodia oblong, falcate; seeds transverse; calyx 5-lobed, hairy, teeth very short ... *Kempeana*

Phyllodia longitudinally striated, without prominent nerves.
   Phyllodia narrow-linear; seeds longitudinal; calyx 5-lobed, hairy, teeth very short ... *doratoxylon*
   Phyllodia linear or narrow-lanceolate, compressed, recurved-pointed, nearly glabrous or hoary; seeds oblique; pod flat, obliquely-oblong; sepals 5, linear-spathulate ... ... *aneura*
      Seeds longitudinal; pod cylindrical, straight ... ... *cibaria*
   Phyllodia linear-subulate, terete, minutely pubescent, 6 to 10 in. long; seeds oblique; calyx 5-cleft ... ... *cyperophylla*
      Phyllodia 2 to 3 in. long; calyx 4-cleft   *Burkittii*

        XII. Leaves all bipinnate.

Stipules spinescent. Pinnae 4 to 6 pairs, leaflets 10-20 pairs ... ... *Farnesiana*

Stipules small or wanting.
   Pinnae 2 to 3 pairs; leaflets 3 to 6 pairs, oblong. Shrub ... ... *Mitchelli*
   Pinnae 8 to 20 pairs; leaflets numerous, close together, several times longer than broad. Trees.
      Pod narrow-linear, much constricted between the seeds; branchlets and foliage with a minute yellowish pubescence ... ... *mollissima*
      Pod broad-linear, hardly constricted between the seeds; branchlets and foliage with a minute whitish pubescence ... ... *dealbata*

ORDER THYMELEAE.
Pimelea.

      I. All flowers with stamens and pistils.

Leaves mostly or all scattered, calyx hairy.
   Flowers in slender interrupted hairy spikes. Erect annual, almost glabrous; bracts 2 or 4 ... ... *trichostachya*

   Flowers in terminal or axillary heads or clusters.
      Involucral bracts 2; flowers very small. Erect, hairy ... ... *curviflora*
      Involucral bracts 4. Erect, glabrous, annual ... *simplex*
      Involucral bracts 6 to 10; leaves erect, concave, oval, midrib prominent. Softly hairy... ... *phylicoides*

> Involucral bracts 8 or more, leaves flat; *heads terminal*. Somewhat shrubby; softly hairy.
>> *Flowers large;* filaments shorter than calyx — *octophylla*
>> Filaments as long as calyx; *bracts numerous* — *petraea*
>
> Leaves mostly or all opposite; involucral bracts 4; flower-heads terminal.
>> Involucral bracts dissimilar to the leaves, shorter than the sepals. Bracts lanceolate-ovate, silky hairy inside, the 2 inner ones much ciliated on the margin; leaves ovate- to narrow-lanceolate, somewhat concave; calyx glabrous at the base. Rather small glabrous shrub ... — *glauca*
>> Involucral bracts unlike the leaves, nearly as long as the sepals.
>>> Bracts beset with silky hairs inside.
>>>> Leaves flaccid, glabrous, oval-lanceolate, about 1 in., flat; flowers large, calyx hairy outside... — *ligustrina*
>>>> Leaves firm, glabrous, about ½ in., incurved at the margin, lateral veins prominent; calyx silky hairy ... — *stricta*
>>> Bracts glabrous on both sides.
>>>> Leaves linear-elliptical, spathulate; calyx hairy throughout ... — *spathulata*
>> Involucral bracts similar to the leaves.
>>> Leaves oval to elliptical, flat, glabrous; branches silky hairy ... — *humilis*

II. Staminate and pistillate flowers on distinct plants.

  a. Leaves opposite, flat; bracts 2 to 4.

> Leaves linear-lanceolate, glabrous; bracts 2 to 4, usually shorter and broader than the leaves; fruit somewhat succulent. Erect, much-branched; calyx beset with short hairs; flowers small, yellow ... — *microcephala*
>
> Leaves under ¼ inch, glabrous, firm, more or less concave; bracts 4, similar to the leaves.
>> Calyx glabrous; a tall, diffuse, glabrous shrub ... — *serpyllifolia*
>> Calyx beset with short and appressed hairs.
>>> Leaves oblong with recurved margins, glabrous. Dwarf ... — *elachantha*
>>> Leaves incurved at the margin, obtuse; calyx-tube of pistillate flower not longer than ovary. Erect ... — *flava*
>>> Leaves flat or concave acute; calyx-tube of pistillate flower produced above the ovary ... — *petrophila*

GENERA AND SPECIES. 81

 b. Leaves scattered; bracts numerous similar to the leaves.

A much branched shrub with oblong or elliptical silky leaves ... ... *ammocharis*

ORDER PROTEACEAE.

 I. Fruit an indeshiscent nut or drupe.

Flowers in dense cone-like spikes, each within a bract; fruit dry; anthers all perfect; ovules 1, rarely 2.
 Bracts firmly adherent to the axis of the cone ... **Petrophila**
 Bracts closely imbricate after flowering, finally falling off with the nut ... ... **Isopogon**
Flowers solitary or few together, each within an involucre of 4 to 8 bracts; fruit dry; anthers all perfect **Adenanthos**
Flowers solitary, axillary, without bracts; fruit a drupe; ovules 2, rarely 1; anthers all perfect ... **Persoonia**
Flowers in short spikes, terminal, each supported by a bract; one of the anthers perfect, 2 imperfect, the fourth abortive; fruit dry, terminated by a tuft of hairs; ovule 1 ... ... **Conospermum**

 II. Fruit bursting by 1 or 2 sutures, 2-seeded.

Fruits distinct and follicular.
 Seeds rarely winged; fruits coriaceous; inflorescence racemose, usually terminal ... ... **Grevillea**
 Seeds with a large black wing; fruits woody; inflorescence usually axillary ... ... **Hakea**
Fruits crowded in dense cones; each fruit compressed, opening at the broad end into 2 hard woody valves. Seeds with a terminal wing ... ... **Banksia**

Petrophila.

A small erect shrub; leaves trichotomously divided into subulate pointed segments; flowers yellow; silky hairy ... ... *multisecta*

Isopogon.

Small erect shrub; leaves ternately or pinnately divided into linear pointed segments; flowers yellow, almost glabrous; involucral bracts longer than the floral bracts ... ... *ceratophyllus*

Adenanthos.

Erect or somewhat prostrate shrubs, very hairy; leaves crowded, rather small and flaccid, of 3 to 7 linear-filiform, erect, segments. Calyx purplish or yellow.

F

Calyx-segments densely bearded inside behind the anthers ... ... *sericea*

Calyx-segments glabrous inside; stem leaves short appressed, the floral ones twice as long ... ... *terminalis*

Conospermum.

Erect, herbaceous, slightly branched above; leaves linear to lanceolate, crowded and very spreading; spikes short on long stalks from the upper axils; flowers blue ... ... *patens*

Persoonia.

A divaricate shrub; leaves linear, pungent-pointed, very spreading; fruit ovate-globular; cotyledons 4 to 6 ... ... *juniperina*

Grevillea.

I. Leaves once or twice dichotomously divided (also *G. pterosperma, G. juncifolia.*)

Leaf-segments broadish-linear, divaricate, pungent; flowers rather large, extremely hairy; racemes short few-flowered; stigma oblique; ovary glabrous stalked ... ... *Huegelii*

Leaf-segments tetragonal subulate; racemes many-flowered ... ... *Treueriana*

II. Leaves simply lobed or sinuate-toothed.

Ovary densely villous, scarcely stalked; stigma slightly oblique.

Leaves ovate-cuneate with broad prickly-pointed angles, silky underneath; style glabrous, scarlet; stigma somewhat oblique, dark-green; racemes dense almost spicate; flowers rather large, hairy outside ... ... *ilicifolia*

Leaves with prickly denticles between the lobes ... *aquifolium*

Ovary glabrous, stalked; stigma oblique.

Leaves ovate or oblong, undulate and prickly toothed, glabrous. Calyx densely bearded inside with erect hairs ... ... *angulata*

Leaves with prickly pointed angles, silky-pubescent; calyx-bearded inside with spreading hairs.

Calyx-tube slightly dilated at the base; leaves ovate ... ... *Wickhami*

Calyx-tube dilated at the base; leaves obovate-cuneate ... ... *agrifolia*

III. Leaves entire.

Leaves linear-terete or almost so, channelled underneath.
    Ovary densely villous, stalked; stigma terminal.
        Leaves 3 to 6 inches, doubly grooved underneath; flowers small; fruit globular; *seeds broadly winged* ... ... *pterosperma*
        Leaves narrow-linear; flowers minute white... *stenobotrya*
    Ovary densely villous; stigma very oblique.
        Leaves 6 to 10 inches, doubly grooved underneath ... ... *juncifolia*
    Ovary glabrous, shortly stalked; stigma terminal.
        Leaves about 1 inch, rigid, pungent-pointed, doubly grooved underneath; flowers very small in very short umbel-like racemes, shortly stalked and terminal... ... ... *halmaturina*
    Ovary glabrous on a long stalk; stigma slightly oblique; leaves obscurely channelled, 3 to 6 inches *mematophylla*

Leaves linear or linear-lanceolate, multistriate; ovary glabrous on a slender stalk; stigma terminal; fruit broad, very oblique, compressed; seeds with an entire wing ... ... *striata*

Leaves flat or with recurved margins. Stigma oblique; calyx bearded inside.
    Racemes terminal umbel-like; style very long; flowers large pink or whitish. Leaves from linear- to elliptical-lanceolate, acute, silky underneath; ovary hairy ... ... *lavandulacea*

    Racemes short, few-flowered, sessile and terminal in the upper axils; style short. Leaves linear, obliquely penniveined and scabrous above ... *aspera*

    Racemes reduced to 1 or 2 pairs of small red flowers, mostly axillary; leaves linear-cuneate; ovary glabrous ... ... *pauciflora*

Hakea.

1. Flowers in cylindrical or spike-like racemes, without involucral bracts (except *H. multistriata*). Small trees.

Leaves terete, very long.
    Racemes glabrous 3 to 4 inches; leaves 6 to 12 inches; stigma broad depressed ... ... *chordophylla*
    Racemes densely hairy; leaves 1 to 2 feet; stigma conical... ... ... *lorea*
Leaves flat, linear-lanceolate, 6 to 12 inches.
    Racemes densely hairy; stigma conical, oblique ... *macrocarpa*

Racemes glabrous, 2 to 3 inches; flowers scarlet; leaves many-streaked, minutely hairy; stigma conical, erect... ... ... *multistriata*

II. Flowers in short racemes or umbel-like clusters enclosed before expansion in imbricating scales. Stigmas depressed and oblique, not conical.

Leaves flat, fan-shaped, prickly-toothed at the margin *Baxteri*

Leaves dichotomously divided into rigid, terete, sharp-pointed segments; flowers yellow, racemose. A small tree with deeply furrowed bark ... ... *Ednieana*

Leaves filiform or linear-terete.
Calyx hairy. Leaves pungent-pointed; flowers in axillary umbels; fruit ovoid somewhat enlarged at the base, compressed and acute at the summit. Small shrub ... ... *vittata*

Calyx glabrous; pedicels pubescent. Leaves compressed filiform about 1 inch; flowers minute, bright yellow in axillary umbels; fruit about 1 inch, obliquely ovate, hardly beaked, smooth or verrucose. Small shrub ... ... *nodosa*

Calyx and pedicels glabrous; fruit ovate.
Flowers in sessile clusters, leaves 3 to 5 inches; fruit rugose, obtuse at the summit with 2 short horn-like excrescences. Small shrub *cycloptera*

Flowers in short racemes, small, white; fruit with a short conical beak. Tall shrub ... *leucoptera*

III. Flowers in short racemes or umbel-like clusters with involucral bracts; stigma conical.

Fruit recurved at the base, terminated in a closely inflexed beak.
An erect shrub of a few feet; leaves terete, pungent-pointed, up to 4 inches long; flowers small white in axillary umbels, pedicels and calyx silky hairy; fruit rugose 1 to 1½ in. long and nearly ¾ in. broad ... ... *rostrata*

A low spreading shrub, leaves and fruit comparatively small ... ... *rugosa*

Fruit with a straight, more or less pointed, apex; calyx and pedicels glabrous.
Leaves dissimilar, pungent-pointed; upper ones linear-trigonous, the lower flat, lanceolate; flowers small white in axillary umbels; fruit obliquely ovate, shortly beaked, smooth or slightly rugose. Tall slender ... ... *ulicina*

Leaves all similar obovate-oblong or -lanceolate, pungent, entire or prickly toothed; flowers small in axillary racemes; fruit broadly ovate, bluntly pointed, with a conical excrescence near the end of one or both valves ... ... *nitida*

Banksia.

Leaves from broad-linear to elliptic-lanceolate, with recurved, entire or slightly denticulated margins, white underneath; style yellowish, at first curved, finally straight; flowers yellow, beset with appressed hairs. Small tree or shrubby ... ... *marginata*

Leaves cuneate-elliptical, regularly serrate, nearly glabrous below; style curved upwards near the base, thence straight and erect; flowers dull-yellow, beset with spreading hairs. Tall shrub ... ... *ornata*

ORDER SAXIFRAGEAE.
Bauera.

Leaves opposite, flowers pink, solitary on slender stalks. Sepals 4 valvate; stamens indefinite. Tallish shrub ... *rubioides*

ORDER CRASSULACEAE.
Tillaea.

Dwarf succulent herbs, leaves opposite, flowers minute, stamens and petals 3 to 5.

Carpels short more or less pointed; flowers axillary.
  Flowers in dense leafy clusters, *sepals* 4 or 5, acute, *longer than the petals*; fruitlets 4 to 5 rather acute, leaves short ... ... *verticillaris*
  Flowers solitary, on long stalks.
    Erect or diffuse of a reddish hue, leaves very short; petals longer than the sepals. ... ... *purpurata*
    Creeping or floating, leaves linear-lanceolate; petals 4 about as long as the sepals; *a scale under each fruitlet* ... ... *recurva*
Carpels oblong; flowers comparatively large, in leafy panicles or corymbs; sepals 4, acute as long as the petals; leaves short, rather acute ... ... *micrantha*

ORDER ROSACEAE.

Petals present; stamens many; carpels indefinite, protruding from the open calyx.
Herbaceous; fruitlets dry; bracts 5.

Styles persistent, calyx-lobes imbricate	**Geum**
Styles deciduous, calyx-lobes valvate	**Potentilla**
Shrubby; fruitlets succulent; no bracts	**Rubus**
Petals absent, stamens not exceeding 10, carpels 1 to 4, enclosed in the calyx-tube.	
Leaves pinnate; ovules pendulous, style terminal; fruit-calyx armed with prickles. Perennial herbs	**Acaena.**
Leaves simple; carpel 1 with 2 erect ovules; stamens 10, hypogynous; style basal. Shrub	**Stylobasium**

Geum.

Erect perennial; radical leaves pinnate of 3 to 5 leaflets on long stalks; flowers yellow, terminal in a loose panicle ... ... *urbanum*

Potentilla.

Creeping; leaves pinnate of several pairs, silvery-white underneath; flowers yellow, solitary on long stalks *anserina*

Rubus.

Erect, lax, prickly stems; leaves pinnate of 3 to 5 leaflets, white-tomentose underneath; flowers reddish, few, in short terminal panicles; fruit red, globular... *parvifolius*

Acaena.

Flowers in cylindrical interrupted spikes, from among almost basal leaves; stamens 5 to 10; fruit-calyx armed with many short barbed prickles ... ... *ovina*

Flowers in globular heads; stems prostrate; stamens 2; fruit-calyx armed with 4 long equal barbed prickles *Sanguisorbae*

Stylobasium.

Erect, leaves cuneate-oblong; drupe nearly dry, globular *spathulatum*

ORDER FICOIDEAE.

Calyx-tube adnate to the ovary.	
Petals numerous linear. Succulent herbs ...	**Mesembrianthemum**
Petals none. Herbs or shrubs	**Tetragonia**
Sepals free from the ovary, but with a distinct tube bearing stamens; petals 0.	
Capsule opening in valves; stamens 4 ...	**Gunnia**
Stamens indefinite; leaves fleshy ...	**Aizoon**
Capsule bursting by a circular rupture.	
Ovary 1-celled, one style	**Trianthema**
Ovary 2-celled; two styles	**Zaleya**
Calyx of distinct sepals; petals 0. Herbs ...	**Mollugo**

Mesembrianthemum.

Leaves sharply triangular, flowers about 2 in. diameter, yellow or pink	*aequilaterale*
Leaves almost cylindrical, bluntly angular; flower about 1 in. diameter, pink	*australe*

Tetragonia.

Prostrate, herbaceous; flowers bisexual; styles 3 or more; fruit with hard protuberances; leaves deltoid, beset with watery papillae ...	*expansa*
Climbing, woody; flowers unisexual, styles 2, fruit succulent red, calyx-lobes yellow inside; leaves lanceolate to ovate-rhomboid... ...	*implexicoma*

Gunnia.

Diffuse annual with opposite linear leaves ...	*septifraga*

Aizoon.

Small rigid shrub, leaves opposite narrow-linear; calyx-lobes ovate, acuminate, whitish inside	*quadrifidum*
Annual, leaves lanceolate-ovate, calyx-lobes profound, lanceolate-oblong, yellow inside ...	*zygophylloides*

Trianthema.

I. Leaves solitary, stalked.

Subshrubby, glabrous; leaves fleshy; stamens 10; capsule ovoid	*turgidifolia*

II. Flowers clustered, axillary, sessile.

Stem prostrate, wiry, glabrous or slightly pubescent, covered with transparent vesicles; stamens 5; capsule short, broad	*crystallinia*
Procumbent, hairy; stamens 20; capsule beaked	*pilosa*
Minute, glabrous; leaves imbricate; capsule globose	*humillima*

Zaleya.

Procumbent, glabrous annual, leaves broadly obovate on long stalks; flowers clustered axillary, stamens 10-12	*decandra*

Mollugo.

I. Stamens and staminodia; seeds with a filiform appendage.

Fertile stamens about 15; flowers large.

Flowers in axillary clusters; usually diffuse coarse plant, starry downy	*hirta*

Flowers in terminal clusters; stout, glabrous *orygioides*
Fertile stamens about 10; flowers small in axillary clusters. Glabrous or slightly pubescent ... *Spergula*

  II. Stamens 3 to 5 all perfect; seeds without appendage.

Glabrous, very small, erect or diffuse; flowers very small on slender pedicels; filaments not dilated ... *Cerviana*

ORDER LYTHRACEAE.

Petals conspicuous; calyx narrow elongated, ribbed, of somewhat herbaceous texture. Tall or dwarf herbs ... **Lythrum**
Petals minute or absent; calyx short membranous.
  Flowers sessile solitary; capsule regularly bursting by valves ... **Rotala**
  Flowers stalked in axillary cymes; capsule bursting irregularly or transversely ... **Ammannia**

Lythrum.

Tall, leaves opposite or whorled; flowers sessile, purplish-red, large, in terminal leafy spikes; stamens 12 ... *Salicaria*
Decumbent annual; leaves alternate; flowers small, axillary, solitary, pink; stamens 6 or less... *hyssopifolia*

Rotala.

Leaves orbicular, opposite; capsule 2-valved; stamens 2 ... *diandra*

Ammannia.

Erect, branching about 2 feet; leaves narrowed at the base; stamens 2 to 4 ... *baccifera*
Erect, branching, dwarf; leaves dilated or cordate at the base; petals 4, minute ... *multiflora*

ORDER ONAGREAE.

Calyx-lobes deciduous; stamens 8; petals 4; seeds hair-tufted **Epilobium**
Calyx-lobes persistent; stamens 10; petals 5; seeds naked **Jussieua**

Epilobium.

Erect, tall herb, glabrous or hairy, more or less 4-angled; leaves opposite or scattered, oblong to linear-lanceolate; flowers pink, axillary, terminal, solitary; fruit filiform-cylindrical ... *glabellum*

GENERA AND SPECIES.

Jussieua.

Herbaceous, creeping in mud or floating in water, bearing cellular floats at the submerged nodes; leaves oval; flowers yellow, axillary, solitary, on long stalks; fruit cylindrical ... *diffusa*

ORDER MYRTACEAE.

I. Ovary 1-celled; fruit not bursting, 1- rarely 2-seeded. Heathy shrubs with small leaves.

Stamens 10, alternating with 10 staminodia.
  Calyx-lobes 5, petal-like, entire ... **Darwinia**
  Calyx-lobes 5, erect, with 3 to 5 hair-like divisions **Verticordia**

Stamens indefinite, in several rows; no staminodia.
  Calyx-lobes with long slender extensions ... **Calycothrix**
  Calyx-lobes truncate or retuse, not pointed ... **Lhotzkya**

Stamens 5 or 10, without staminodia, regularly alternate with or opposite to the calyx-lobes ... **Thryptomene**

II. Ovary 2- to 5- or more-celled; fruit opening by valves.

Stamens 20 or more, in a single row, shorter than petals.
  Leaves opposite; flowers axillary, small; stamens about 20; fruit usually 3-celled. Heath-like glabrous shrubs ... **Baeckea**
  Leaves scattered; flowers solitary, but crowded at the end of the branchlets; stamens numerous, fruit 5- to 10-celled. Erect bushy shrubs **Leptospermum**

Stamens indefinite, exceeding the petals.
Stamens free in more than one row.
  Flowers in dense terminal heads; calyx-lobes persistent; stamens slightly protruding; seeds pendulous. Prostrate shrub ... **Kunzea**
  Flowers in spikes, crowned by the year's shoot; calyx-lobes persistent; stamens much protruding; seeds erect. Tall shrubs with erect branches ... **Callistemon**

Stamens connate in 5 bundles, opposite the petals; ovules several in a cell ... **Melaleuca**

Stamens free in several rows; petals absent; calyx truncate crowned in the bud by a cap or lid *(operculum)*, formed of the united calyx-lobes; flowers in umbels or in panicles ... **Eucalyptus**

Darwinia.

Flowers in small globular heads; leaves very small, linear, 3-angled; calyx-lobes longer than the petals; ovules 4. Erect ... ... *micropetala*

Fowers solitary, axillary; leaves small shortly acute; calyx-lobes as long as petals; ovules 2. Prostrate *Schuermanni*

Verticordia.

Erect, bushy; flowers small in umbel-like corymbs... *Wilhelmii*

Calycothrix.

Calyx-tube 6 lines long; the upper free portion as long and not more slender than the adnate portion, the lobes short and broad with long hair-like awns. Petals 4 to 5 lines ... ... *longiflora*

Calyx-tube 2 to 4 lines, the lower fusiform portion joined by a long slender column to the bell-shaped free portion; petals pink or whitish, 2 lines long; leaves triangular or quadrangular, linear ... ... *tetragona*

Lhotzka.

Calyx-tube very narrow-turbinate; whole plant glabrous; bracteoles shorter than calyx; leaves spreading, 1½ lines long ... ... *glaberrima*

Calyx-tube cylindrical; plant pubescent; bracteoles as long as calyx.
Leaves spreading 1 to 2 lines; calyx-tube hirsute atop... ... ... *genetylloides*
Leaves rather appressed, 1½ lines; calyx-tube glabrous, constricted at the summit ... ... *Smeatoniana*

Thryptomene.

I. Calyx-tube rugulose, not ribbed.

Calyx-tube hemispheric, lobes triangular with scarious basal extensions; ovules 4 to 6 ... ... *Maisonneuvii*

Calyx-tube ovate-obconic, lobes semiorbicular, yellow, ciliate; stamens alternate; ovules 8 to 12 ... *flaviflora*

II. Calyx-tube not rugulose, with or without ribs.

Calyx-tube broad, compressed, ovoid-bellshaped, almost smooth; lobes longer than the petals; stamens 5 ... ... *Mitchelliana*

Calyx-tube not ribbed, semiovate; lobes reniform-cordate, margins petaloid, auricled at the base... *auriculata*

Calyx-tube ribbed, truncate-ovate ... ... *Elliotti*

Calyx-tube 10-ribbed, cylindrical; stamens 5.
    Leaves three-cornered; flowers solitary, small, on very short axillary stalks near the summit of the branchlets ... ... *ericaea*
    Leaves flat or slightly concave, oblong ... ... *Miqueliana*
Calyx-tube 5-ribbed, ovate-turbinate; lobes shorter than the petals; leaves obovate, triangled, thick, obtuse, generally ciliated; stamens 5; ovules 4   *ciliata*

Baeckea.

    I. Stamens 10 to 30, of which 5 are opposite the centre of the petals; filaments filiform; ovary convex atop.

Stamens 10; ovules 2, 3, or rarely 4, in each cell.
    Leaves linear, spreading; flowers large, solitary, pink, on stalks much longer than the leaves. Subshrubby with long, lax, more or less prostrate branches ... ... *diffusa*
    Leaves thick, linear-terete; flowers small, solitary, white, on stalks shorter than the leaves. Dwarf erect shrub ... ... *crassifolia*
Stamens 15; ovules 2 in each cell; leaves imbricate; flowers sessile, solitary very small; bracteoles broad, white or reddish ... ... *ericaea*
Stamens 20 to 30; ovules 8 in a cell; leaves linear-terete ... ... *polystemona*

    II. Stamens 8 to 15, none opposite the centre of the petals; filaments clavate under the anther; ovary flat-topped.

Leaves rather long, linear-cylindrical with a recurved point; flowers small, white, solitary; fruit 3-celled ... ... *Behrii*

Leptospermum.

    I. Fruit usually 10-celled, flat atop; perfect seed flat with a membranous margin.

Flowers large sessile; calyx glabrous; leaves obovate-oblong, glabrous ... ... *laevigatum*

    II. Fruit usually 5-celled, convex atop; perfect seed narrow-linear.

Calyx-tube glabrous; leaves lanceolate, pointed ... *scoparium*
Calyx-tube pubescent.
    Leaves obovate ½ inch long, more or less silky; calyx-lobes as long as the tube ... ... *lanigerum*
    Leaves smaller, clothed with appressed shining hairs; calyx-lobes exceedingly short ... ... *myrsinoides*

Kunzea.

Rigid, prostrate; leaves cordate to ovate-orbicular; flowers white, sessile, forming dense terminal heads; fruit-calyx succulent; fruit globular, 3-celled. Mostly maritime ... ... *pomifera*

Callistemon.

Leaves lanceolate, thick, flat; filaments red ... ... *coccineus*

Leaves lanceolate or linear, flat; filaments greenish-yellow ... ... *salignus*

Leaves linear-subulate, terete; filaments red.
   Leaves above 2 in. long; filaments above ½ in., hairy ... ... *teretifolius*
   Leaves under 1½ in., sharp-pointed; filaments short ... ... *brachyandrus*

Melaleuca.

I. Filaments red or purplish in small heads or clusters, or short spikes.

Leaves scattered, small, ovate- to linear-lanceolate; flowers in globular heads; filaments purplish or occasionally white ... ... *squamea*

Leaves opposite, narrow; calyx-lobes herbaceous, persistent; flowers few together; claws of the staminal bundles about twice as long as the petals; fruit immersed in the rhachis ... ... *Wilsoni*

Leaves opposite, broad; calyx-lobes scarious and deciduous; flowers in heads or short spikes; claws of the staminal bundles very short; fruits immersed in the thickened rhachis.
   Leaves oval or obovate, rarely ¼ in. long ... *gibbosa*
   Leaves oblong-lanceolate to broad-linear, under ½ in. long, in four decussate rows on the branchlets ... ... *decussata*

II. Filaments yellow; flowers in terminal spikes or heads. Tall shrubs.

Leaves opposite, ovate-cordate to-lanceolate, ½ in. long, 5- to 7-nerved; flowers in oblong or cylindrical spikes ... ... *squarrosa*

Leaves alternate linear to linear-subulate, 1 to 2 in. long; flowers in globular heads ... ... *glomerata*

III. Filaments white or whitish (also *M. squamea*).

Spikes elongate-cylindrical.
   Leaves mostly opposite, linear or linear-lanceolate. Tall tree ... ... *trichostachya*

Leaves scattered, fruits immersed in the rhachis.
  Stamens longer than the petals; leaves rather short, linear- to narrow-lanceolate, almost flat, recurved or spreading; calyx-lobes deciduous. Small tree, bark persistent; orifice of fruit lobeless ... ... *parviflora*
  Stamens 3 or 4 times longer than the petals, the filaments pinnately arranged on the stalk of the staminal bundles; leaves semi-terete, very narrow, under $\tfrac{1}{2}$ in.; orifice of fruit permanently lobed... ... *cylindrica*

Spikes globular or short.
  Leaves elongate, recurved-pointed.
    Leaves opposite, narrow- to broad-lanceolate, 3 to 4 lines long; flowers in small clusters; fruits nearly globular, truncate, immersed *acuminata*
    Leaves opposite in 4 decussate rows; flowers in well-developed heads ... ... *quadrifaria*
    Leaves scattered, linear-subulate, 1 or 2 in. long; heads globular. Very tall, glabrous shrub ... ... *uncinata*
  Leaves short and blunt, glandular-rough, nerveless.
    Leaves narrow-linear; flowers in short spikes. Tall shrub; freshwater swamps chiefly *ericifolia*
    Leaves oblong-linear; flowers in small terminal leafy heads. A very tall shrub or small tree with a thin papery bark peeling off in layers. Salt marshes ... ... *pustulata*

Eucalyptus.

  I. Fruit cylindrical-ovate, about twice as long as wide.
Fruit slightly urn-shaped, ribbed, valves enclosed; lid hemispheric; umbels solitary; anthers long, opening by parallel slits. Tall tree with smooth bark ... ... *corynocalyx*
Fruit not ribbed; valves enclosed, lid patellar; umbels solitary or partly paniculate; anthers roundish, opening by parallel slits. Shrub; leaves opposite, broad, connate ... ... *gamophylla*
    II. Fruit truncate-ovate longer than wide, base narrowed.
Fruit somewhat semi-cylindrical, about half as long again as wide; valves enclosed.
  Fruit slightly urceolate; umbels paniculate; leaves narrow-elongate ... ... *tessellaris*

Fruit ribbed, *stalks broadly compressed;* umbels solitary ... *incrassata*

Fruit smooth, lid semiovate-conical; calyces somewhat angular; umbels paniculate ... *hemiphloia*

Fruit more or less obconic and urceolate, lid hemispheric; calyces *ribbed;* umbels solitary. Small shrub ... *gracilis*

Fruit urceolate, smooth, *annular at the edge;* lid hemispheric; umbels solitary ... *odorata*

Fruit truncate-ovate, a little longer than wide.

    a. Anthers kidney-shaped opening by divergent slits; umbels solitary; border of fruit-orifice depressed.

Leaves elongate, *veined longitudinally;* lid hemispheric; pedicels very short ... *pauciflora*

Leaves copiously *pellucid-dotted,* veins not spreading; lid almost hemispheric. Here, shrubby ... *amygdalina*

Leaves very inequilateral at the base; calyces granular-rough; lid hemispheric; pedicels very short. Tall tree, bark persistent ... *obliqua*

Leaf-veins not much spreading, pedicels compressed; lid hemispheric or patellar ... *Sieberiana*

    b. Anthers roundish opening by minute pores (also *E. odorata* and *E. hemiphloia); umbels paniculate.*

Leaves paler beneath; lid conic-semiovate ... *paniculata*

Leaves equally dull-green; lid double, the inner one hemispheric, the outer smaller ... *largiflorens*

Leaves thick, broadish, equally green, shining; flowers small; lid hemispheric ... *Behriana*

    c. Anthers roundish opening by longitudinal slits; umbels solitary.

Leaves thick; *stalks compressed,* thick; lid nearly hemispheric; fruit ribbed ... *incrassata*

Fruit-valves long-pointed, *exsert;* stalks slender, stalklets very short; lid conical pointed ... *oleosa*

    d. Anthers longer than broad, opening by longitudinal slits.

Leaves thick dull-green; fruit urceolate, *lid irregularly separating;* umbels paniculate ... *terminalis*

Leaves narrow-elongate, fruit slightly urceolate, lid patellar, pedicels very short, *bark with intersecting fissures,* umbels paniculate ... *tessellaris*

GENERA AND SPECIES.

Umbel solitary, *stalk compressed; fruit* 1- *to* 4-*angled*, valves usually 3 or 4, hardly exsert, lid pyramidal-hemispheric ... ... *goniocalyx*

III. Fruit semiovate to semiglobose, about as long as wide, base rounded.

    a. Anthers roundish, opening by pores.

Umbels solitary mostly 3-flowered, pedicels elongate, flowers white, sometimes pink or scarlet; fruit with 1 more or less prominent rib, lid conic-hemispheric. Timber-tree, bark deciduous ... *leucoxylon*

Umbels solitary, pedicels very short ; *leaves* narrow-lanceolate, *dark-dotted;* lid semiovate or narrow-conical ; *fruit-valves exsert*... ... ... *uncinata*

    b. Anthers roundish, opening by slits.

Leaves dark-green, *narrow-elongate;* umbels solitary, peduncles very short, pedicels 0 ; lid semiovate; fruit-valves slightly exsert... ... ... *cneorifolia*

Leaves dull and pale-green ; umbels paniculate ; fruit somewhat obconic, lid semiovate-conical, valves much exsert ... ... *microtheca*

    c. Anthers longer than broad, opening by slits.

Fruit convex at the summit; valves exsert ; umbels solitary.

    Umbels few-flowered, pedicels very short, lid nearly hemispheric ; fruit-valves very small ... *Stuartiana*

    Umbels mostly 3-flowered, pedicels 0, or very short ; *lid semiovate*, short-pointed ... ... *viminalis*

    Umbels several-flowered, peduncles elongate, *pedicels conspicuous; lid elongate, sharp-pointed*   *rostrata*

Fruit flat or depressed atop (also *E. capitellata*).

    Umbels solitary, several-flowered, pedicels very short ; lid hemispheric, short-pointed ; valves slightly exsert ... ... *Gunnii*

    Umbels solitary, peduncles very short, pedicels 0; lid almost hemispheric, short-pointed; fruit comparatively very large, valves exsert ... *cosmophylla*

    IV. Fruit more or less biconic, the dorsal portion hemispheric; valves exsert; umbels solitary.

    a. Upper portion of fruit obtusely conical, truncate. Anthers kidney-shaped, opening by divergent slits.

Pedicels almost wanting ; lid semiovate-conical. Shrub ... ... *santalifolia*

Pedicels wanting, lid, hemispheric. Tall tree, bark persistent ... *capitellata*

Pedicels short, lid concavely attenuated, sharply-pointed. Tall tree with persistent outer bark... *macrorrhyncha*

Peduncles and pedicels short; lid semiovate hemispheric ... *Oldfieldii*

 b. Upper portion of fruit acutely conical, truncate; umbels solitary. Fruits very large.

Leaves very thick broadish; basal half of fruit 4-angled; lid ridged; filaments yellow; anthers round ... *pachyphylla*

Leaves thick, ovate, basal-half of fruit longitudinally wrinkled; lid rugose; filaments red; anthers longer ... *pyriformis*

ORDER RHAMNACEAE.

Calyx spreading, petals absent; fruit 1-celled, 1-seeded, produced into an oblong terminal wing ... **Ventilago**

Calyx campanulate or tubular, *adnate to the fruit*.
 Petals 0 or minute, not enclosing the large oblong anthers. Calyx-tube entirely adnate; bracts deciduous ... **Pomaderris**
 Petals minute enclosing the short small anthers. Calyx-tube extended beyond the fruit; bracts persistent ... **Cryptandra**

Ventilago.

Leaves lanceolate, flowers panicled; small glabrous tree *viminalis*

Pomaderris.

i. Petals very narrow.

Branchlets and underside of leaves silky; leaves broadly oblong or obovate, small; calyx-tube very short, silky *mrytilloides*

ii. Petals absent.

Flowers numerous in much-branched panicles; calyx starry-hairy; leaves ovate-lanceolate, acute, crenate, wrinkled on upper side, 2 to 4 in. long. Tall shrub *apetala*

Panicles raceme-like; calyx large, lobes persistent, tube very short; leaves ovate to orbicular, thick, rust-coloured underneath, crenate or entire. Small erect shrub ... *racemosa*

Panicles small crowded, leaves obcordate or 2-lobed, white underneath; calyx-tube turbinate. Low shrub *obcordata*

Cryptandra.

1. Bracts small; *flowers stalked (Trymalium).*

Leaves obovate-spathulate, glabrous above, grey-velvety below; panicles short, few-flowered. Tall lax shrub ... *Wayii*

11. *Flowers sessile, densely crowded,* surrounded by small persistent imbricate brown bracts *with 1 or more floral leaves (Spyridium).*

a. Leaves ovate to orbicular.

Leaves from obovate to orbicular, wrinkled; calyx-tube very hairy, short, the lobes glabrous; floral leaves roundish, white-tomentose. Branches slender erect, softly hairy ... *Hookeri*

Leaves obovate to ovate, glabrous above, with raised, dense, *reticulations;* bracts pubescent ... *phlebophylla*

Leaves cuneate-obovate about ½ in., glabrous above, silky below; bracts minute; calyx very short ... *spathulata*

Leaves obovate or ovate, distinctly stalked, very obtuse or emarginate, densely tomentose on both sides; calyx slender, very hispid; floral leaves several, all woolly-white ... *coactilifolia*

Leaves obovate or obcordate with a recurved point, shortly stalked, white-tomentose above, rust-coloured with appressed hairs below; floral leaves white; calyx silky-hairy, the tube long and slender; bracts orbicular to lanceolate. Rather dwarf and spreading ... *leucophracta*

Leaves obovate or oblong, obtuse, smooth above, tomentose below; calyx-tube short hairy, the lobes glabrous; bracts orbicular. Low, much-branched, beset with a rusty tomentum ... *obovata*

b. Leaves linear to elliptical, entire, revolute at the margin.

Clusters of flowers stalked; floral leaves ovate and shortly stalked, or broadly linear and sessile, white-tomentose on both sides; calyx short hispid; leaves glabrous above, stipules small. Prostrate or suberect ... *vexillifera*

Clusters of flowers sessile; floral leaves usually one; leaves tomentose on both sides, stipules on young shoots large; calyx hirsute or tomentose ... *subochreata*

c. Leaves narrow-cuneate, 2-lobed, or notched at the summit, the margins recurved. Much-branched, moderately tall shrubs.

Leaves cuneate-oblong, lobes short blunt, densely stellately-hairy, underside also with long simple hairs ... *halmaturina*

Leaves narrow-cuneate, lobes longer subacute, upper side glabrous or nearly so, underside densely beset with long hairs ... ... *bifida*

Leaves linear-oblong, somewhat clustered, deeply channelled above, bluntly notched at the end; upper side scabrous and sparsely hispid, underside with long subappressed hairs ... ... *scabrida*

III. *Flowers sessile in loose cymes* surrounded by small persistent imbricate brown bracts and 1 or 2 floral leaves.

Erect, *viscid*. Leaves narrow-linear, margins revolute, glabrous above; stipules linear-lanceolate; calyx-tube narrow-turbinate beset with spreading hairs... ... *Waterhousei*

IV. *Flowers* sessile or shortly stalked, *in* clusters or *leafy spikes*, never in cymes, surrounded by persistent imbricate brown bracts, *and each flower by brown bracteoles*. Leaves small, narrow, revolute.

Flowers closely sessile in small terminal or lateral clusters; calyx silky-hairy rather large and broad, the *lobes short and spreading;* free part of the ovary longer than the adnate portion. Leaves linear-terete; *bracteoles acuminate* and ciliate, longer than the calyx *hispidula*

Flowers in small terminal clusters; *calyx about ½ in.*, silky-hairy, the *lobes as long as the tube*. Leaves clustered; bracteoles almost rhomboid, ciliate; branchlets somewhat spinescent, nearly glabrous ... *propinqua*

Flowers in short leafy spikes; calyx broadly campanulate, tomentose, the lobes shorter than the tube. Leaves narrow-elliptical to roundish-ovate, somewhat recurved at the margin or almost flat; bracteoles obtuse shorter than the calyx ... ... *amara*

Flowers crowded into clusters; *calyx glabrous, the lobes as long as the tube;* leaves revolute; branchlets tomentose ... ... *tomentosa*

ORDER OLACINEAE.

Olax.

Calyx enlarged after flowering and enclosing but free from the fruit, staminodia 2-cleft. Glabrous shrub. Leaves thick obovate ... ... *Benthamiana*

GENERA AND SPECIES.

ORDER SANTALACEAE.

I. Calyx-tube adnate to the ovary.

Leaves comparatively large, usually opposite; anthers opening by longitudinal slits. Shrubs or small trees; flowers paniculate; fruit a drupe ... ... **Santalum**

Leaves small or minute, scattered; anthers opening by two or four lobes. Shrubs.
Each flower surrounded by 2 to 4 scaly bracts; anthers 4-lobed... ... ... **Choretrum**
Each flower subtended by one bract; anthers 2-lobed **Leptomeria**

II. Calyx free from the ovary.

Flowers unisexual; pistillate flowers solitary, staminate flowers clustered ... ... **Anthobolus**
Flowers bisexual in spikes; leaves rudimentary; fruit-stalks succulent ... ... **Exocarpos**

Santalum.

I. Calyx-tube adnate at the base, upper part free from the ovary.

Small tree with pendulous branches; leaves lanceolate, long; fruit small, elliptical, black; stigma 3- to 4-lobed ... ... *lanceolatum*

II. Calyx-tube wholly adnate *(Fusanus)*; fruit globose.

Endocarp of fruit deeply wrinkled; mesocarp succulent, sweet, bright-red; leaves narrow-lanceolate, acuminate... ... ... *acuminatum*
Endocarp slightly pitted; mesocarp hardly succulent, bitter, brownish-red; leaves linear to narrow-lanceolate acute ... ... *persicarium*

Choretrum.

Flowers in clusters of 2 to 5; clusters shortly stalked and arranged in racemes; branchlets lax, angular.
Corolla white, deeply cleft ... ... *glomeratum*
Corolla yellow, deeply cleft ... ... *chrysanthum*
Flowers solitary sessile, but crowded into rather long spikes; branchlets rigid, striated ... ... *spicatum*

Leptomeria.

Erect, branches spinescent, flowers in short spikes ... *aphylla*

Anthobolus.

Leafless furrowed branches; fruit-stalks conspicuous ... *exocarpoides*

Exocarpos.

I. Spikes cylindrical, usually shortly pedunculate.

Leaves minute, scale-like; a small tree with erect branchlets; fruit-stalks bright-red, very succulent, longer than broad... ... ... *cupressiformis*

Leaves linear-subulate; a small tree with pendent branchlets; fruit-stalks almost dry, green ... ... *spartea*

II. Spikes very short and scarcely pedunculate, the rhachis pubescent.

A shrub with robust hardly angular branchlets; leaves in form of scales; fruit-stalks very succulent, bright red ... ... *aphylla*

III. Spikes reduced to sessile clusters of 2 or few flowers.

A shrub with angular branchlets; leaves minute, linear; fruit-stalks very succulent, pale-lilac, or whitish ... *stricta*

ORDER HALORAGEAE.

Flowers with calyx; petals present at least in the males. Fruit usually 1-seeded; flowers in corymbose panicles; petals 2 to 4, large, yellow; stamens 4 to 8; stigmas 2 to 5. Somewhat shrubby, erect herbs of a somewhat bluish tinge, with scattered, narrow, entire leaves... ... ... **Loudonia**

Fruit 2- to 4-seeded, not separating into fruitlets; flowers solitary, or clustered within each bract forming a simple or paniculate terminal raceme; petals and stigmas 4, rarely 2 or 3, small; stamens twice as many as petals. Herbs, or somewhat shrubby ... ... **Haloragis**

Fruit of 2 or 4 separable nut-like fruitlets; flowers very small, axillary, the upper ones usually males, the lower ones females. Aquatic, the submerged leaves usually capillary-divided ... ... **Myriophyllum**

Flowers without calyx and corolla, within two bracteoles; stamen 1; styles 2; ovary 4-celled with 1 ovule in each cell. Aquatic or mud plants with opposite simple leaves ... ... **Callitriche**

Loudonia.

Petals usually 2; stamens 4, styles 2, fruit 2-winged ... *Behrii*

Petals 4; stamens 8, styles 4, fruit with 4 broad wings *aurea*

Haloragis.

i. Calyx-lobes, petals and pistils constantly 2 (*Meionectes*).

Semiaquatic, glabrous; leaves scattered, pinnatisect; flowers few, axillary ... ... *Meionectes*

ii. Calyx-lobes, petals and pistils 4, or rarely 3 or 2.

Leaves alternate, the lower ones digitately lobed ... *heterophylla*
Leaves alternate, narrow-linear, entire.
    Leaves semiterete, rather fleshy; glabrous; styles and ovules 2 or 4; fruit ovoid-globular, smooth... *digyna*
    Leaves recurved along the margin; densely hairy; fruit ovoid, rough; styles and ovules 4 ... ... *elata*
Leaves alternate, linear or lanceolate, toothed or pinnatifid.
    Fruit ovoid-globular or somewhat quadrangular, often rugose or muricate; leaves nearly sessile, coarsely toothed; flowers 1 or 2, axillary ... *aspera*
    Fruit acutely 4-angled ... ... *acutangula*
    Fruit quadrangular, large, the basal and apical angles with tooth-like excrescences; leaves stalked; flowers clustered ... ... *odontocarpa*
    Fruit acutely trigonal, ovate; leaves stalked, remotely serrate, lanceolate. Erect, glabrous ... *trigonocarpa*
    Fruit winged, 3-celled; leaves shortly stalked ... *Gossei*
Leaves opposite; flowers solitary, axillary.
    Prostrate, glabrous; bracts minute; leaves orbicular-cordate, closely serrate; fruit minute, globular, prominently 8-nerved ... ... *micrantha*
    Erect, scabrous or hirsute; leaves short, distantly serrate; fruit wrinkled and rough, globular-quadrangular.
        Leaves ovate- to narrow-lanceolate; upper bracts minute ... ... *tetragyna*
        Leaves ovate- to orbicular-cordate; upper bracts larger ... ... *teucrioides*

Myriophyllum.

i. Leaves all entire.

Leaves alternate, very small, linear-cylindrical; stamens 2 or 4 ... ... *integrifolium*
Leaves opposite; stamens 8. Small creeping mud-plants.
    Leaves oblong; fruitlets 4, smooth ... ... *amphibium*
    Leaves linear; fruitlets 4, tuberculate; staminate flowers generally stalked ... ... *pedunculatum*

II. Leaves all capillary-pinnatisected.

Leaves whorled; emerged leaves pinnatifid; fruitlets rough ... *verrucosum*

Leaves opposite; fruitlets smooth; male flowers enclosed before expansion in a petaloid hood-shaped bract ... *Muelleri*

III. Emerged leaves entire; submerged leaves divided into long capillary segments.

Emerged leaves narrow-linear; calyx-lobes conspicuous ... *intermedium*

Emerged leaves oval or broadly lanceolate; calyx-lobes minute ... *elatinoides*

Callitriche.

Mostly submerged; upper leaves obovate to spathulate, lower ones linear ... *verna*

ORDER UMBELLIFERAE.

Fruit ovate of a single carpel, 1-seeded; styles 2 ... **Actinotus**

Fruit of two cohering fruitlets, ulimately separating, each provided with a style and 1-seeded.
  Fruitlets more or less laterally compressed, without oil-ducts.
    Umbels simple; fruitlets often only moderately compressed; involucral bracts few or absent; stipules scarious ... **Hydrocotyle**
    Umbels simple; fruitlets flat or much compressed; involucral bracts many, connate below; exstipulate ... **Didiscus**
    Umbels compound, stipules absent.
      Calyx-teeth inconspicuous; fruit slightly compressed ... **Trachymene**
      Calyx-lobes peltate; fruit much compressed **Xanthosia**
    Flowers in dense spikes or heads, surrounded by rigid sharp-pointed bracts; flowers blue; fruitlets scarcely compressed; leaves radical elongate, prickly ... **Eryngium**
  Fruitlets slightly or not compressed; oil-ducts present.
    Fruitlets with 5 prominent ribs, almost or quite smooth.
      Umbels simple; leaves linear-terete... **Crantzia**
      Umbels simple; leaves much divided **Caldasia**
      Umbels compound; calyx-teeth inconspicuous; one oil-duct under each furrow ... **Apium**
        Calyx-teeth prominent; several oil-ducts under each furrow ... **Sium**

Fruitlets with 4 prominent bristly ribs; umbels simple ... *Daucus*

Actinotus.

Tufted, tall, hairy; leaves cleft into several narrow segments; umbels on long stalks ... *Schwarzii*

Hydrocotyle.

I. Leaves without lobes. Perennials, glabrous, creeping and rooting.

Leaves peltate, orbicular; petals minute, pink, valvate; semiaquatic ... *vulgaris*
Leaves broadly cordate; petals broad, imbricate. Marsh plant ... *Asiatica*

II. Leaves lobed or segmented. Perennials with creeping rooting stems; petals greenish or yellowish.

Leaves reniform-cordate with 3, 5 or more, short blunt crenate lobes.
    Flowers numerous, more or less unisexual, in each umbel; male flowers stalked, female flowers almost sessile; fruitlets nearly truncate. Softly hairy ... *Candollei*
    Flowers few and sessile in each umbel.
        Fruitlets slightly angular. Softly hairy ... *hirta*
        Fruitlets with an expanded dorsal angle. Glabrous ... *pterocarpa*
Leaves divided to the middle into 3 or 5 crenate lobes, small; glabrous; fruitlets crowned with flattened bristly hairs ... *comocarpa*
Leaves cleft to near the base into 3 or 5 small wedge-shaped segments. Slender slightly hairy ... *tripartita*

III. Leaves more or less deeply cleft. Small or filiform, erect or diffuse annuals, not rooting at the nodes.

Fruitlets not much compressed.
    Fruitlets smooth with 3 obtuse semicircular ribs on the back, a well-defined pit on each side. Minute, tufted, glabrous ... *callicarpa*
    Fruitlets smooth with 3 obtuse semi-circular ribs on the back; the sides tuberculate, with a well-defined pit. Prostrate and diffuse elongate stems.
        Stems filiform; leaves divided below the middle ... *trachycarpa*
        Stems stout; leaves divided to near the base ... *crassiuscula*
Fruitlets pitted and rugose on the back, the sides smooth. Minute slender, glabrous ... *capillaris*

Fruitlets with 3 thick obtuse wings, very rugose between them. Minute, slender, glabrous ... *medicaginoides*

Fruitlets very flat, when ripe with a very prominent dorsal rib. Small, slender, diffuse, glabrous ... *diantha*

Didiscus.

I. Small annuals; flowers few in the umbels.

One of the fruitlets smooth or granulated, the other prickly-rough. Leaves small, narrowly lobed; petals white ... *pusillus*

Fruitlets equally beset with bristles; leaves small, narrowly lobed; petals blue ... *cyanopetalus*

Fruitlets covered with dense cottony wool ... *eriocarpus*

II. Coarse erect plants, flowers in large umbels; leaves divided.

More or less hirsute, petals white ... *pilosus*

Glabrous and glaucous, petals blue ... *glaucifolius*

Trachymene.

Leaves narrow-linear; the lower ones cleft into linear lobes, or entire. Herbaceous, slender, branched, erect; fruitlets granular-rough... ... *heterophylla*

Xanthosia.

Leaves cleft into three elliptical or lanceolate segments; umbels 1- to 4-flowered. A dwarf hairy perennial ... *pusilla*

Leaves cleft into three narrow subdivided segments; umbels irregularly compound. Dwarf, glabrous ... *dissecta*

Eryngium.

Flower-heads ovoid or globular.
  Stems erect; leaves doubly or simply pinnatifid; spinular bracts exceeding the heads ... *rostratum*
  Stems prostrate; leaves simply pinnatifid; spinular bracts much exceeding the heads ... *vesiculosum*
Flower-heads oblong-cylindrical; leaves long, linear; spinular bracts chiefly shorter than the heads ... *plantagineum*

Crantzia.

Stems slender, creeping and rooting at the nodes; leaves linear-terete ... *lineata*

Caldasia.

Erect, slightly branched; leaves repeatedly pinnately divided ... *andicola*

Apium.

Stems prostrate or decumbent, rarely erect; segments of leaves from broad-linear to rhomboidal ... ... *prostratum*

Sium.

Large, erect; leaves simply pinnate; umbels terminal ... *latifolium*

Daucus.

Small erect annual beset with short stiff hairs; umbels of few very unequal rays; leaves twice pinnate of incised segments ... ... *brachiatus*

Sub-Class III.—Synpetaleae Perigynae.

ORDER CUCURBITACEAE.

i. Anther-cells very flexuous or conduplicate.

Fruit with a hard rind; connective produced beyond the anthers; petals almost distinct; calyx tubular and campanulate, segments subulate ... ... **Cucumis**

Anthers without an appendage; peduncles all slender, 1-flowered, with a cordate bract below the flower; corolla 5-partite ... ... **Momordica**

ii. Anther-cells straight, parallel.

Stigmas 3-fringed; corolla campanulate, ciliate or toothed; calyx 5-toothed ... ... **Melothria**

Cucumis.

Very villous; stems flexuose, pentagonal; leaves rotund, obtusely angular, toothed; fruit elliptical, hairy ... *Chate*

Momordica.

Leaves palmately 7-lobed, dentate, somewhat hairy; tendrils downy; fruit oblong-acuminate, angular, tuberculate; seeds stalked ... ... *Charantia*

Melothria.

Leaves orbicular-cordate, palmately 5- to 7-lobed; male and female flowers in the same axils, both minute and shortly stalked; fruit small, globular, smooth ... *Muelleri*

Leaves deeply cordate or hastate with broad rounded or angular lobes, obscurely crenate; male flowers sessile *Maderaspatana*

ORDER LORANTHACEAE.

Flowers bisexual; anthers bursting longitudinally ... **Loranthus**
Flowers unisexual; anthers opening by pores ... ... **Viscum**

Loranthus.

I. Anthers versatile; petals free, yellowish ... *celastroides*
II. Anthers adnate; petals united to the middle.

Flowers in cymes; leaves linear opposite ... ... *angustifolius*
Flowers solitary or in pairs; petals usually 6.
    Leaves flat, opposite; petals scarlet, green at the summit; fruit orange, turning to red and finally purple ... ... *Exocarpi*
    Leaves filiform-cylindrical.
        Leaves mostly opposite; pedicels terete; berries white, globular; petals scarlet... ... ... *linearifolius*
        Leaves alternate; pedicels shortly winged; petals pale yellow, rose-coloured at the summit *Murrayi*

III. Anthers adnate; petals free.

Flowers in axillary cymes.
    Leaves terete.
        Leaves glabrous; flowers usually glabrous, the common peduncles of the cyme very short, bearing 3 or 4 rays of 3 or rarely 5 flowers each; petals usually 5 ... ... *linophyllus*
        Leaves hoary-tomentose, flowers tomentose, the common peduncle of the cyme bearing 2 rays of 2 pedicellate flowers each, or the cyme reduced to a pair of pedicellate flowers; calyx-tube gibbous at the base by reason of the adnate bracteole; petals 4 ... ... *gibberulus*
    Leaves flat.
        Flowers all stalked, or the central ones sessile; petals 5, foliage and inflorescence usually glabrous ... ... *pendulus*
        Flowers sessile, petals green; berry green with a thick epicarp; foliage and inflorescence hoary *Quandang*
Flowers sessile between two large bracts. ... ... *grandibracteus*

Viscum.

Leaves absent; branches flattened; petals 3, minute, persistent ... ... *articulatum*

ORDER RUBIACEAE.

I. Leaves opposite; ovules several in each cell; fruit capsular scarcely dehiscent. Perennial herbs.

Corolla bell-shaped, entire; sepals and petals 4 ... ... **Oldenlandia**
Corolla toothed; sepals and petals 5 ... ... **Dentella**

GENERA AND SPECIES.

  II. Leaves opposite; ovule 1 in each cell.

Fruit a berry-like drupe. Shrubs.
 Flowers in axillary cymes; ovule laterally attached   **Canthium**
 Flowers in small terminal heads; ovule erect   ...   **Coprosma**

Fruit dry, capsular, 2-valved; 1 ovule in each cell.
 Fruits connate in heads; ovule erect. Herbs   ...   **Opercularia**
 Fruits connate in umbels; ovule erect   ...   ...   **Pomax**
 Fruits in heads, but not connate; ovule laterally attached ... ...   **Spermacocce**

  III. Leaves whorled, rarely reduced to one pair; fruit dry, 2-lobed, indehiscent, 1 ovule in each cell. Weak herbs with quadrangular stems.

Corolla funnel-shaped with a distinct tube, at least of the staminate flowers; flowers more or less unisexual ...   **Asperula**
Corolla rotate without any conspicuous tube   ...   ...   **Galium**

Oldenlandia.
Dwarf perennial; leaves linear; stipules small, toothed   *tillaeacea*

Dentella.
Prostrate or creeping herb, flowers solitary; stipules scarious   ...   .   *repens*

Canthium.
Glabrous shrub; leaves broadly ovate, very rigid   ...   *latifolium*

Coprosma.
Rigid shrub; leaves ovate, shortly acuminate, scabrous above   ...   .   *hirtella*

Opercularia.
Peduncles erect; leaves linear; seeds obtusely angled.
 Usually erect, scabrous-pubescent; heads globular on long stalks   ...   .   *scabrida*
Peduncles recurved; seeds broad.
 Small procumbent; leaves ovate; seeds smooth, furrowed along the inner side; stamens 3 to 4   ...   *ovata*
 Diffuse or wiry; leaves small, oblong-lanceolate or almost linear; seeds somewhat wrinkled, with 2 prominent ribs on the inner face; stamens 2   ...   *varia*

Pomax.
Somewhat shrubby and dwarf, more or less hairy; leaves ovate to lanceolate   ...   .   *umbellata*

Spermacocce.

Prostrate; leaves sessile, ovate to broad-lanceolate, with callous margins ... *marginata*

Asperula.

Leaves linear in pairs; stems very slender; flowers minute ... *geminifolia*

Leaves linear to oval, 4 to 8 in a whorl; flowers in terminal clusters ... *oligantha*

Galium.

Fruit glabrous and smooth; leaves 4 in a whorl, narrow; flowers white, axillary crowded, peduncles short ... *umbrosum*

Fruit rough with hooked bristles; leaves 4 in a whorl, narrow-lanceolate to ovate; peduncles elongated .. *australe*

CAPRIFOLIACEAE.

Sambucus.

Stems herbaceous, erect, 3 to 5 feet; flowers white in wide corymbs; corolla-lobes and stamens 3, rarely 4; berries white. ... *Gaudichaudiana*

ORDER COMPOSITAE.

I. Ray-flowers ligulate in one row; disk-flowers tubular, *leaves opposite*, very small bracts between the flowers.

Phyllaries in 2 rows, the outer ones narrow, leafy and glandular; the inner ones nearly ovate. No pappus ... **Siegesbeckia**

Phyllaries in 2 or 3 rows almost equal.

Pappus reduced to 1 to 4 minute teeth or short bristles ... **Wedelia**

Pappus of 2 to 4 rough spines.

Ray-flowers sterile, achenes quadrangular... **Bidens**

Ray-flowers fertile, achenes flattened ... **Glossogyne**

Phyllaries few; flower-heads small, narrow, collected in dense clusters surrounded by floral leaves; pappus 0 ... **Flaveria**

II. *Ray-flowers ligulate;* disk-flowers tubular, leaves alternate or radical.

a. Pappus of capillary bristles.

Ray-flowers in one row, achenes cylindrical.

Pappus-bristles unequal in 1 or 2 rows, anthers obtuse.

Phyllaries in several rows; stigmas flattened, papillose ... **Aster**
Phyllaries in 1 row; stigmas truncated, hair-tufted ... **Senecio**
Pappus-bristles denticulate, anthers with basal points ... **Pterigeron**
Ray-flowers in two or more rows, blue. Achenes produced into a slender beak ... **Podocoma**
Achenes compressed, not beaked; pappus-bristles unequal, in more than two rows ... **Vittadinia**

  b. Pappus of capillary bristles and scales. Ray-flowers and phyllaries in 2 or more rows.

Fertile achenes with several rows of capillary bristles and a few lanceolate scales; sterile achenes with bristles only; ray white ... **Dimorphocoma**
Fertile achenes compressed, with capillary bristles; sterile achenes with scales ... **Minuria**

  c. Pappus of lanceolate-subulate flat segments.

Achenes angular; ray-flowers in one row; bracts between the flowers... **Achnophora**

  d. Pappus of rigid spines (also Glossogyne).

Achenes angular; ray-flowers in one row; phyllaries in 2 rows ... **Calotis**

  e. Pappus very short or wanting. Ray-flowers in 1 row.

Achenes compressed, cylindrically produced, glandular hairy; phyllaries in several rows. Ray blue **Lagenophora**
Achenes various, truncated; phyllaries in 2 rows, nearly equal, margins membranous ... **Brachycome**
Achenes oblong, papillose; phyllaries in 1 row; bracts between the flowers. Ray blue ... **Erodiophyllum**
Achenes oblong, glabrous, ribbed on the back and sides, phyllaries in several rows. Ray yellow... **Cymbonotus**

  III. *Flowers all tubular*, or the marginal flowers not conspicuously rayed, in distinct, not compound heads. Female flowers usually outside, in one or more rows, and more slender than the inner bisexual ones.

   a. Leaves radical or tufted, phyllaries in two rows; anthers obtuse at the base.

Pappus wanting; achenes ovate-angular, glabrous ... **Solenogyne**

Pappus of oblong scales, achenes silky-hairy; heads
clustered ... ... **Isoetopsis**

b. Leaves scattered.

Phyllaries in one row.
    Pappus absent; achenes pointed. Slender dwarf
    annuals ... ... **Toxanthus**
    Pappus of lanceolate scales attenuated into long
    points ... ... **Quinetia**
    Pappus of ciliated or plumose bristles, or wanting; achenes beaked. Slender annuals ... **Millotia**
    Pappus of capillary bristles.
        Marginal flowers slender, pistillate, in 2 or
        3 rows. Erect coarse herbs ... ... **Erechthites**
        Flowers all tubular and bisexual (in some
        species) ... ... **Senecio**

Phyllaries in two or more rows.
    Pappus wanting; anthers obtuse at the base.
        Corolla of marginal flowers undeveloped,
        or minute; achenes flattened, heads
        stalked. Dwarf herbs ... ... **Cotula**
        Corolla of marginal flowers minute, tubular;
        achenes triangular or quadrangular, heads
        sessile ... ... **Centipeda**
        Achenes flat, bordered by herbaceous wings **Ceratogyne**
        Flowers unisexual on distinct plants; fertile
        achenes without pappus; sterile achenes
        with flattened ciliolate bristles ... ... **Ethuliopsis**
    Pappus wanting; anthers with basal points.
        No scales between the flowers.
            Phyllaries herbaceous or scarious; heads
            axillary sessile. Herbs ... ... **Epaltes**
            Inner phyllaries with recurved points... **Stuartina**
            Phyllaries scarious, appressed. Erect,
            glabrous sticky shrubs; heads in compact corymbs ... ... **Humea**
        Receptacle-scales present. Inner phyllaries
        white, radiating; outer ones appressed,
        glutinous ... ... **Ixodia**
    Pappus of chaffy scales.
        Scales narrow-lanceolate; anthers obtuse at
        the base ... ... **Elachanthus**
        Scales spathulate or obovate; anthers with
        basal points ... ... **Rutidosis**
    Pappus of capillary bristles, simple, denticulate
    or plumose.

(1) *Phyllaries herbaceous* (also *Aster tubuliflorus* and *A. axillaris*).

Pappus of simple capillary bristles; style simple ... **Pluchea**

Pappus-bristles denticulate; style bulbous at the base... ... ... **Pterigeron**

Pappus-bristles flattened, plumose; achenes stalked ... ... **Podosperma**

(2) *Phyllaries*, at least the inner ones, with short scarious radiating tips.

Pappus-bristles barbed; phyllaries narrow ... **Ixiolaena**

Pappus-bristles plumose; phyllaries with long points ... ... **Athrixia**

(3) *Phyllaries scarious appressed*.

Receptacle-scales present; pappus-bristles simple or denticulate ... ... **Cassinia**

Receptacle-scales absent; pappus-bristles simple. Phyllaries all thin and scale-like; marginal flowers in 1 row, much enlarged, sometimes ligulate. Erect herbs ... ... **Podolepis**

    Marginal flowers in 2 rows, achenes blunt. Woolly annuals. ... ... **Gnaphalium**

    Outer phyllaries scarious, the inner ones only at the tips; achenes narrowed upwards ... ... **Leptorrhynchos**

(4) *Inner phyllaries petal-like and spreading*.

Pappus-bristles plumose from the base ... ... **Helipterum**

Pappus-bristles simple or denticulate.

    Achenes beaked; pappus-bristles scabrous **Waitzia**

    Achenes truncated; bristles simple... ... **Helichrysum**

  IV. *Flower-heads clustered within a general involucre;* the partial heads also with involucral bracts. Leaves alternate.

    a. Partial heads without scales between the flowers.

General involucre of many bracts in several rows.

  Bracts of general involucre with large white appendages forming rays; pappus-bristles ciliate or plumose... ... ... **Polycalymma**

  Bracts without appendages; pappus rudimentary ... ... **Hyalolepis**

General involucre wanting or of a few bracts (also *Isoetopsis, Flaveria* and *Stuartina*).

  Partial heads 1- to 3-flowered, their phyllaries *few, compressed*, and scarious; pappus wanting

or cup-shaped or rarely of plumose scales. Dwarf herbs more or less cottony.
 Compound heads on an elongate receptacle **Anglanthus**
 Compound heads on a flat receptacle ... **Skirrophorus**
Partial heads 1- or 2-flowered, their phyllaries several, the outer ones narrow, the inner ones broader; pappus various ... ... **Gnephosis**
Phyllaries of general and partial involucres more or less scarious; pappus-bristles plumose ... **Calocephalus**
Partial heads 1-flowered enveloped in wool; outer phyllaries of the general involucre herbaceous, the inner ones with broad scarious margins; pappus wanting or of capillary bristles ... ... **Eriochlamys**
Inner phyllaries with long, petaloid, radiating laminae; pappus-bristles with a plumose tuft subtended by a scale ... ... **Cephalipterum**
Compound heads surrounded by crowded leaves; pappus of 5 plumose bristles ... ... **Gnaphalodes**

  b. Partial heads with receptacle-scales.

Compound heads elongate or globular; bracts of general involucre woolly, of the partial heads linear and scarious; pappus of simple bristles united in a ring at the base. Coarse, woolly, herbaceous perennials; leaves decurrent ... **Pterocaulon**
Compound heads ovoid or globular; bracts of general involucre with brown margins and more or less concealed; partial heads 3- to 8-flowered, phyllaries scarious; pappus of plumose bristles. Erect herbs, woolly or silky ... ... **Craspedia**
Compound heads ovoid-globular, sessile within radical leaves; pappus wanting ... ... **Chthonocephalus**

  v. *Flowers all ligulate;* leaves radical.

Phyllaries in 2 rows; pappus of linear scales, fine-pointed ... ... **Microseris**

Siegesbeckia.

Rather tall, pubescent; leaves ovate-triangular, heads in leafy panicles ... ... *orientalis*

Wedelia.

Prostrate or ascending beset with appressed hairs; leaves lanceolate, sessile; pappus wanting or of 4 minute teeth; heads axillary, stalked ... ... *platyglossa*
Erect, scabrous; leaves oblong-lanceolate, stalked; pappus cup-shaped ... ... *verbesinoides*

Bidens.
Glabrous perennial; leaves bipinnate; ray yellow ... *bipinnata*

Glossogyne.
Glabrous perennial; leaves alternate, pinnate; achenes striate ... *tenuifolia*

Flaveria
Glabrous annual; leaves linear or linear-lanceolate; ray yellow ... *Australasica*

Aster.
1. Vestiture of underside of leaves silky, cottony, or woolly. Ray white.

Leaves scattered.
 Heads large solitary on long stalks or terminating long branchlets.
  Leaves ovate 2 to 4 in., toothed, silky below   *Sonderi*
  Leaves ovate 2 to 4 in., entire, densely cottony below ... *pannosus*
  Leaves oblong-cuneate, about ¼ in., recurved at the margin, hoary above, tomentose below   *pimeloides*
 Heads small in terminal leafy racemes or panicles.
  Leaves obovate to oblong or lanceolate, toothed, silky below ... *myrsinoides*
  Leaves narrow-linear, recurved margins, woolly below ... *Mitchelli*

Leaves clustered, small, rarely over ¼ inch, recurved at the margin; heads comparatively small, achenes hairy.
 Ligule of ray-flowers small or almost obliterated, shorter than the style; heads very small, axillary, sessile or nearly so.
  Flowers few in each head; leaves linear, ¼ to 1 inch ... *tubuliflorus*
  Flowers 10 to 15 in each head; leaves obovate-cuneate to linear, ½ in. ... *axillaris*
 Ligule of ray-flowers developed, longer than the style. Heads terminating short axillary branchlets, but more or less crowded in leafy paniculate spikes.
  Flowers about 20 in each head.
   Leaves orbicular- to elliptic-ovate, 1 to 2 lines long ... *microphyllus*
   Leaves almost linear, to about ½ in. long   *ramulosus*
   Leaves obovate, mostly 3-toothed, stalked   *exiguifolius*

Flowers less than 10 in each head; leaves minute. Leaves appressed, branchlets with a woolly tomentum ... ... *lepidophyllus*

   II. Vestiture of underside of leaves consisting of stellate hairs; achenes hairy.

Heads rather small, stalked in leafy panicles; ray-florets about 10, white; leaves ovate-oblong or lanceolate, large; involucre turbinate ... ... *stellulatus*

Heads larger, stalked, solitary and terminal; ray-florets about 20, blue; leaves oblong-linear, sinuate-toothed; involucre hemispheric ... ... *asterotrichus*

   III. Glabrous; branchlets and foliage sticky, Ray white, rarely blue.

Leaves nearly or quite flat.
  Heads singly terminal; leaves small more or less toothed.
    Heads very large, leaves narrow-cuneate, 3-toothed at the end; achenes glabrous ... *magniflorus*
    Leaves obovate; achenes silky hairy ... *calcareus*
   Heads rather small; leaves stalked, toothed.
    Leaves obovate or broadly cuneate, under ½ inch ... ... *Muelleri*
    Leaves narrow-oblong or-lanceolate, under 1 inch ... ... *Stuartii*
  Heads in leafy panicles, comparatively small.
    Leaves decurrent on the branchlets, linear-oblong, entire or coarsely toothed, ½ to 1 inch long, achenes silky hairy ... ... *decurrens*
    Leaves simply sessile, narrow-linear, somewhat revolute, ½ to 1½ inch long; achenes sparingly hairy ... ... *glutescens*
Leaves linear-cylindrical, closely revolute.
  Leaves appressed, under ¼ in. long, crowded; heads quite small, singly terminating leafy branchlets; ray bluish ... ... *teretifolius*
  Leaves spreading, long, acute, glandular-dotted; heads small in terminal leafless corymbose-panicles ... ... *glandulosus*

   IV. Leaves glabrous or sprinkled with rigid simple or jointed hairs. Heads large; rays blue or purplish.

Leaves obovate-cuneate, stalked, deeply indented; heads solitary, slenderly stalked; fruits compressed, sparingly silky-hairy ... ... *megalodontus*

Leaves broadly lanceolate, remotely toothed, stalked;
heads few in a terminal corymb; peduncles as long
as the leaves with subulate bracts; fruits silky ... *Ferresii*

Leaves obovate-oblong or oblong-cuneate, toothed,
sessile; heads solitary or few in a terminal
corymb; fruits glabrous, striate ... ... *exul*

Leaves linear, acute, sessile; heads solitary on long
slender stalks; fruits glabrous or silky ... ... *Huegelii*

Podocoma.

Much-branched; leaves acutely-toothed, cuneate, beset
with long rigid hairs ... ... *cuneifolia*

Vittadinia.

Dwarf, erect, branches leafy; leaves obovate or spathulate to linear-cuneate, entire or indented, hairy;
achenes finely striate ... ... *australis*

Dimorphocoma.

Minute, erect, beset with jointed hairs, slightly
branched; leaves flat, entire... ... ... *minutula*

Minuria.

I. Corolla of ray-flowers blue.

Achenes of ray-flowers, silky hairy; phyllaries oblong-
linear, margins scarious.
   Low, branching undershrub; often almost glabrous; leaves narrow-linear; heads terminal,
   stalked ... ... *leptophylla*

Achenes of ray-flowers almost glabrous; phyllaries
narrow, acute.
   Ligule of ray-corollas elongated; involucre about
   3 lines long.
      Rather tall glabrous, undershrub, leaves cylin-
      dric-linear, entire and acute; heads com-
      paratively large on terminal peduncles ... *Cunninghamii*
   Ligule of ray-corollas short; involucre under 2
   lines long.
      Leaves lanceolate or linear, very acute; glau-
      cous ... ... *integerrima*
      Leaves linear, obtuse, toothed; branchlets
      woolly ... ... *denticulata*

II. Corollas of ray-flowers yellow.

Achenes of ray-flowers glabrous; phyllaries oblong-
linear; pappus-bristles united in a tube at the base.
A small glabrous undershrub, leaves cylindric-
linear; heads very small on short stalks ... ... *suaedifolia*

Achnophora.

Leaves linear, radical; heads single terminating long stalks; ray blue ... ... *Tatei*

Calotis.

I. Pappus consisting of spines and scales.

Stems leafy; leaves cuneate or spathulate toothed at the end, with a clasping base; pappus-spines 2 or 3, slender, barbed, alternating with broad scales. Ray-flowers long and narrow, blue ... ... *cuneifolia*

Stems prostrate; leaves ovate to lanceolate, stalked; pappus-spines 4 to 8, short, barbed, mostly alternating with cleft scales. Ray yellow ... ... *hispidula*

II. Pappus of spines only, united at the base.

Pappus-spines two. Branches ascending; pubescent or hirsute; leaves linear-cuneate or oblong, coarsely toothed; ray yellow ... ... *cymbacantha*

Pappus of 3 to 5 spines. Plant erect, glabrous; leaves linear or linear-lanceolate, acute, entire; ray yellow *erinacea*

III. Pappus of several unequal spines, distinct.

Achenes not winged.
  Leaves radical, scapes simple; stems creeping.
    Plant hirsute, leaves obovate or oblong, toothed or lobed; pappus-spines about 8. Ray blue ... ... *scabiosifolia*
    Plant glabrous; leaves linear entire. Ray yellow ... ... *scapigera*
  Stems erect, branched, leafy; pappus-spines 4 to 8 barbed.
    Stem-leaves linear; pappus longer than the achene ... ... *lappulacea*
    Stem-leaves cuneate; pappus shorter than the achenes ... ... *microcephala*
Achenes bordered by densely ciliate wings. Small erect annuals.
  Achenes covered by plumose hairs; pappus-spines several, unequal, about as long as the achene; ray white ... ... *plumulifera*
  Achenes shortly hirsute; pappus-spines numerous, shorter than the achene. Ray purple ... ... *porphyroglossa*

IV. Pappus membranous, annular, without spines.

Erect, glandular-downy, leaves narrow-lanceolate, serrate; ray yellow ... ... *Kempei*

Lagenophora.

Herbs; leaves radical obovate to cuneate-oblong, toothed or indented; heads solitary on long stalks. Root-stock *emitting* slender *rhizomes;* phyllaries narrow ... ... *Billardieri*

Plant larger; leaves coarser and more indented; phyllaries rather broad ... ... *Huegelii*

Brachycome.

I. Ray inconspicuous; pappus conspicuous.

Dwarf; branches leafy; leaves linear, lobed or toothed; achenes angular ... ... *goniocarpa*

Leaves radical, linear, 3-lobed; scapes leafless, 1-headed; achenes compressed, ciliate on the sides ... ... *pachyptera*

II. Ray inconspicuous; pappus absent.

Dwarf glabrous annual; leaves linear pinnatifid; achenes flat, bordered by a wing cleft into hooked lobes ... ... *collina*

III. Ray conspicuous; pappus absent or rudimentary. (Glabrous perennials with bluish ray-flowers.

Stem decumbent, leaves pinnatifid; achenes glabrous with thick obtuse margins. ... ... *Muelleri*

Leaves linear entire, achenes compressed, with thick margins, the sides often rough ... ... *graminea*

Stem erect, achenes narrow, margins thick, sides tuberculate.

Rigid; leaves broadish, entire ... ... *basaltica*

Branches slender, rather spreading; leaves narrow, the lower ones lobed ... ... *trachycarpa*

IV. Ray conspicuous; pappus conspicuous.

Stems tall simple, slightly leafy; heads large solitary and terminal; ray white; radical leaves pinnatifid, hirsute; achene obliquely elliptical, hardly compressed, somewhat angular ... ... *diversifolia*

Stems branched and leafy.
  Achenes bordered by a wing.
    Ray-flowers blue; leaves pinnate or lobed.
      Achene-wing ciliated; plant glabrous or woolly ... ... *ciliaris*
      Plant hirsute, dwarf; leaves pinnatifid or trifid ... ... *debilis*
    Ray yellow; leaves linear, entire, glandular-hairy ... ... *chrysoglossa*

Ray white or pink; leaves simple or toothed; achenes granular-rough on the sides. Robust, clothed with white wool ... ... *calocarpa*
Achenes with thickened margins, not winged.
Achenes brown quadrangular; leaves deeply cleft into narrow lobes. Dwarf, glandular-hairy ... ... *exilis*
Achenes black, sides tuberculate; leaves oblong-cuneate, toothed. Rather dwarf; glandular-hairy ... ... *melanocarpa*
Leaves radical; scapes leafless; glabrous herbs; ray blue.
Achenes winged.
Leaves linear, entire; heads large on stalks about 1 ft. ... ... *cardiocarpa*
Leaves narrow-cuneate, spathulate, toothed. Annual ... ... *cuneifolia*
Achenes with thickened margins, not winged.
Leaves obovate-oblong, toothed towards the summit ... ... *decipiens*
Leaves pinnatifid, segments linear; achenes ciliate on the sides ... ... *pachyptera*

Erodiophyllum.

Hispid; leaves pinnately lobed, stalked; peduncles 1-headed... ... ... *Elderi*

Cymbonotus.

Leaves radical, ovate, toothed or lobed; scapes very short ... ... *Lawsonianus*

Solenogyne.

Leaves elliptic-cuneate, 2 to 3 in.; scapes robust about as long as the leaves ... ... *Emphysopus*

Isoetopsis.

Dwarf, leaves linear; heads in sessile compound clusters *graminifolia*

Toxanthus.

Phyllaries recurved at the tips, achenes much pointed; leaves linear. More or less densely vested with long woolly hairs ... ... *perpusillus*
Phyllaries erect; achenes slightly pointed, sparsely and minutely hairy. More or less extensively beset with glandular-hairs ... ... *Muelleri*

Quinetia.

Dwarf, erect, woolly tomentose; leaves linear-cuneate to obovate, stalked ... ... *Urvillei*

Millotia.

Stems erect; white with close or woolly hairs; pappus-bristles as long as the corolla, scarcely ciliolate; corolla pale; leaves narrow-linear ... ... *tenuifolia*

Prostrate; pappus-bristles shorter than the corolla, ciliate plumose ... ... *Greevesii*

Erect, woolly; pappus absent; corolla yellow ... ... *Kempei*

Erechthites.

I. Phyllaries not exceeding 12.

Involucres relatively short, not exceeding 3 lines long.
  Heads in a loose panicle; phyllaries 8 to 10.
    Nearly glabrous, leaves lanceolate, regularly toothed ... ... *prenanthoides*
    Scabrous, leaves coarsely lobed... ... ... *picridioides*
  Heads in contracted panicles; phyllaries about 12.
    Hispid; leaves lanceolate, toothed or lobed; achenes short, minutely hairy ... ... *arguta*

Involucres relatively long, 4 lines long, phyllaries about 12; heads in a loose, sometimes crowded, panicle.
  Scabrous; leaves deeply lobed or divided, cottony below ... ... *mixta*
  Vestiture cottony; leaves linear, entire; achenes rather long ... ... *quadridentata*

II. Phyllaries 15 to 20; involucre 4 to 5 lines long, broad.

Leaves linear or lanceolate, entire or coarsely indented or lobed, cottony underneath ... ... *hispidula*

Senecio.

I. Ray-flowers well developed.

Erect leafy annuals, glabrous or almost so; heads large.
  Heads solitary, terminal; phyllaries united to above the middle; leaves entire ... ... *Gregorii*
  Heads few in a corymb; leaves pinnatifid ... ... *platylepis*

Glabrous undershrubs, heads large in leafy corymbs.
  Leaves all toothed; upper ones clasping; involucre with a few accessory bracts. ... ... *spathulatus*
  Upper leaves entire, stem-clasping; accessory bracts inconspicuous.
    Leaves oblong; heads about 3 inches diameter *megaglossus*

Leaves obovate, upper ones with basal lobes;
heads smaller ... ... *magnificus*

Erect perennial herbs.
Heads not exceeding 1 in. diameter in corymbs; leaves linear-lanceolate, entire, or sometimes toothed; involucre campanulate, Glabrous ... *lautus*

Heads comparatively small; involucre cylindrical.
Heads few, hoary; leaves linear, simply sessile ... ... *Behrianus*

Heads numerous, glabrous; leaves frequently dilated at the base; achenes usually glabrous ... ... *dryadeus*

II. Flowers all tubular; or the ligule of the ray-flowers rudimentary; involucre cylindrical. Shrubs, except, *S. brachyglossus*.

Leaves pinnate, segments long linear. Glabrous ... *anethifolius*

Leaves oblong or lanceolate, dilated, auricled and stem-clasping; glabrous ... ... *odoratus*

Leaves simple, narrowed into a petiole.
Leaves ovate, white below ... ... *hypoleucus*

Leaves linear. Usually glabrous ... ... *Cunninghamii*

Leaves linear or lanceolate; cottony below ... *Georgianus*

Erect leafy annual, rarely branched; leaves linear or narrow-lanceolate, entire or indented. Heads very small; ray-corollas with minute ligules ... ... *brachyglossus*

Cotula.

I. Receptacle flat or convex; marginal flowers without corolla.

Achenes of marginal flowers in a single row; leaves entire.
Phyllaries very broad; achenes of disk-flowers, not winged ... ... *filifolia*

Phyllaries ovate; disk-achenes winged; leaves sheathing ... ... *coronopifolia*

Achenes of marginal flowers in several rows; stems slender prostrate; leaves dissected; heads on long stalks ... ... *australis*

II. Receptacle conical; marginal flowers with a short corolla.

Glabrous creeping perennial; leaves dissected; phyllaries orbicular ... ... *reptans*

Centipeda.

Prostrate, lax; leaves entire, not dilated at the base; heads shortly stalked; achenes ellipsoid-clavate, striated ... *orbicularis*

Erect, rigid; leaves toothed, dilated at the base; heads sessile.
   Heads almost hemispherical; marginal flowers in many rows; achenes cylindric-clavate ... *Cunninghamii*
   Heads almost semiovate; marginal flowers in few rows; fruit cylindrical, striated ... *thespidioides*

Ceratogyne.

Dwarf erect annual; leaves stalked; heads sessile axillary ... *obionoides*

Ethuliopsis.

Erect, glabrous, branched; heads terminal and loosely clustered... *Cunninghamii*

Epaltes.

Diffuse perennial; phyllaries orbicular, herbaceous; leaves obovate to spathular-cuneate, stalked; achenes glabrous ... *australis*

Dwarf annual; phyllaries oval, scarious, ciliate; leaves oblong-lanceolate, sessile; achenes scabrous ... *Tatei*

Stuartina.

Diffuse slender annual; leaves orbicular, stalked; flower-heads very small in little globular clusters, sessile among floral leaves ... *Muelleri*

Humea.

Leaves crowded, scale-like; flowers yellow, 4 or 5 within each head; achenes rough ... *squamata*

Leaves linear, semiterete, obtuse, clustered; flowers white, 3 within each head; achenes glabrous ... *cassiniacea*

Ixodia.[1]

Erect, glabrous, sticky shrub; leaves linear-lanceolate, sessile or decurrent on the stem; heads crowded in corymbs ... *achilleoides*

Elachanthus.

Slender, dwarf, branching annual; leaves small linear; heads terminal; fertile achenes obovate, silky-hairy ... *pusillus*

Rutidosis.

Erect, cottony, perennial; leaves linear, sessile; pappus-scales 5 to 7; flowers yellow; heads on terminal peduncles ... ... *helichrysoides*

Minute diffuse annual; heads in a dense terminal cyme *Pumilio*

Pluchea.

Heads cylindrical, singly terminal; flowers 4 or 5 in each head, phyllaries blunt. Erect woolly shrub, leaves small ... ... *conocephala*

Heads ovoid in corymbose clusters; flowers several; inner phyllaries acute, rigid. Glandular-pubescent perennial ... ... *tetranthera*

Heads broadly ovoid in small corymbs; flowers several; inner phyllaries narrow long-pointed. Glabrous perennial ... ... *Eyrea*

Pterigeron.

Heads large, hemispheric; leaves obovate, narrowed at the base, to oblong-cuneate, toothed. *Ray-flowers ligulate.* Erect, scabrous ... ... *liatroides*

Heads ovoid; leaves oblong, narrowed at the base ... *microglossus*

Heads narrow-ovoid; leaves linear, stalked ... ... *adscendens*

Heads ovate-companulate; leaves obovate, crenate-toothed ... ... *dentatifolius*

Podosperma.

Dwarf ascending; leaves linear; involucre cylindrical, at length conical, 1 to 2 inches long; achenes hairy *angustifolium*

Ixiolaena.

I. Involucres hemispheric, phyllaries very narrow; pappus nearly as long as corolla.

Leaves lanceolate, acute, narrow at the base; heads on long stalks with distant bracts; phyllaries glandular-hairy; pappus bristles 8 to 12 ... ... *leptolepis*

Leaves oblong-spathulate, stalked; peduncles rather short with small bracts; phyllaries scabrous; pappus-bristles 20 to 30 ... ... *supina*

II. Involucre companulate, pappus as long as corolla.

Leaves lanceolate, stem-clasping; phyllaries woolly ... *tomentosa*

Athrixia.

Dwarf pubescent annual; heads singly terminating long stalks ... ... *tenella*

Cassinia.

I. Small erect shrubs with cylindric-linear leaves about 1 inch.

Leaves scabrous above; phyllaries dull-white; flowers within each headlet 6 to 12; heads in dense corymbs ... *aculeata*

Leaves white-tomentose beneath; headlets with 3 to 5 flowers.
  Panicles long and loose; leaves short, recurved-pointed; phyllaries transparent shining... ... *arcuata*
  Panicle not longer than broad, leaves long; phyllaries white ... ... *laevis*

Leaves glabrous, keeled, acute; corymbs small; one flower in each headlet; branchlets sticky; phyllaries pale-yellow ... ... *punctulata*

II. Very tall erect herb with large flaccid leaves.

Heads in large loose terminal panicles; headlets with many flowers ... ... *spectabilis*

Podolepis.

I. Heads ovoid-cylindric.

Phyllaries yellow, wrinkled, acute; leaves lanceolate or ovate-lanceolate ... ... *rutidochlamys*

II. Heads hemispherical, exceeding ½ inch diameter.

Phyllaries smooth, acute, yellow.
  Annual beset with cottony hairs; heads under 1 in. diameter ... ... *canescens*
  Perennial, nearly glabrous; heads over 1 inch diameter ... ... *acuminata*

Phyllaries wrinkled, obtuse; heads relatively large. Stout, perennial, glabrous, or beset with loose cottony hairs ... ... *rugata*

II. Heads almost hemispherical, under ½ inch diameter; phyllaries smooth, yellow. Erect annuals with filiform branches.

Leaves ovate to lanceolate, small, stem-clasping, cottony hairy; all the flowers of about equal length ... ... *Lessoni*

Leaves linear, glabrous; marginal flowers longer, whitish ... ... *Siemessenia*

Gnaphalium.

Clusters of heads terminal, leafless; leaves flaccid; densely woolly ... ... *luteo-album*

Clusters of heads axillary or in terminal leafy spikes.
Loosely woolly ... ... *Indicum*

Clusters of heads terminal, surrounded by floral leaves; leaves firm, bright-green above. Beset with somewhat cottony or appressed hairs ... ... *Japonicum*

Dwarf, beset with dense cottony wool; clusters of heads surrounded by narrow leaves; leaves flaccid *indutum*

Leptorrhynchos.

I. Phyllaries acute.

Pappus-bristles not exceeding 10; achenes very shortly contracted atop; phyllaries minute, ciliate. Heads on long stalks.
  Pistillate flowers without pappus, the others with 4 to 6 bristles plumose at the end; leaves linear revolute ... ... *tenuifolius*
  All the flowers with a pappus.
    Bristles of male flowers 8 to 12, denticulate; of the others 3 to 5; leaves lanceolate. Perennial scantily beset with cottony wool *squamatus*
    Bristles of male flowers 4, denticulate; of the others 2 or 3. Leaves linear-lanceolate. Annual beset with some cottony wool ... *pulchellus*

Pappus-bristles numerous; outer phyllaries glabrous, inner ones ciliate; heads on long bracteate stalks. Beset with scattered scale-like hairs.
  Achenes glandular-rough, elongated. Perennial... *elongatus*
  Achenes smooth, with a long beak. Annual ... *medius*

II. Phyllaries broad, obtuse, ciliate; achenes beaked.

Robust annual, somewhat cottony; pappus-bristles numerous ... ... *Waitzia*

Helipterum.

I. Inner phyllaries with ray-like tips.

Stems and foliage glabrous (also *H. laeve*). Heads solitary on long stalks.
  Involucre hemispheric.
    Pappus plumose at the end. Ray pink or white *roseum*
    Pappus equally plumose from the base.
      Ray white; leaves linear, often crowded.. *anthemoides*
      Ray yellow; leaves oblong-spathulate ... *polygalifolium*
  Involucre ovoid; ray white, leaves mostly clasping *strictum*

Stems and foliage scantily beset with hairs or nearly glabrous.

Involucre hemispheric; heads solitary on leafless stalks.
    Achenes glabrous, flat; ray yellow; leaves narrow-linear, short. Dwarf erect annual; pappus-bristles 8 to 12, yellow at the tips ... *hyalospermum*
    Achenes silky; leaves stalked, oblong or lanceolate; scabrous-pubescent somewhat viscid annual ... ... *heteranthum*
    Achenes silky; heads sometimes paniculate; ray white; pappus-bristles 7 to 10 ... ... *floribundum*
Involucre somewhat turbinate or cylindrical, achenes silky; leaves narrow-linear or filiform; heads sessile in dense corymbs.
    Ray yellow or white; bristles 15 to 20 ... *tenellum*
    Ray white, bristles numerous ... ... *pygmaeum*
Stems or foliage or both woolly or cottony invested; leaves linear.
    Achenes silky; heads in corymbs; ray white ... *corymbiflorum*
    Achenes glabrous, smooth or papillary.
        Inner phyllaries attenuated into stalks, heads solitary on long leafless stalks.
            Outer phyllaries subulate; ray yellow; leaves stem-clasping ... ... *stipitatum*
            Outer phyllaries broad; ray white or pinkish; leaves crowded at the base ... *incanum*
        Inner phyllaries with broad claws.
            Achenes rough; heads solitary; ray white or yellow ... ... *Cotula*
            Achenes smooth; heads in corymbs; *leaves decurrent* ... ... *Haigii*

II. Inner phyllaries without ray-like tips.

Stems and foliage glabrous. Phyllaries brown; achene glabrous ... ... *laeve*
Stems and foliage scantily hairy or nearly glabrous.
    Phyllaries rigid; achene rough; pappus of marginal flowers wanting, of the others of 1 to 4 flat ciliate bristles. Dwarf ... ... *dimorpholepis*
    Phyllaries membranous; achene smooth; pappus-bristles 10, plumose. Minute ... ... *exiguum*
Stems or foliage or both, woolly or cottony invested. Heads small, usually in dense terminal corymbose clusters.
    Achenes almost glabrous; flowers in each head 2 to 5 ... ... *moschatum*
    Achenes papillary; flowers 15 to 20; pappus-bristles 8 to 5 ... ... *pterochaetum*

Achenes scantily silky; flowers and bristles 10 to 12 ... *Tietkensi*
Achenes silky; heads in spike-like panicles, flowers 7 to 13; pappus-bristles, 15 to 20... ... ... *Charleysae*

Waitzia

Stem erect, simple, herbaceous, somewhat hairy. Heads in corymbs; phyllaries yellow, acute, ciliate; leaves linear, long ... ... *corymbosa*

Helichrysum.

1. Inner phyllaries with ray-like tips.

Marginal flowers fertile, their achenes hairy; inner flowers sterile, their achenes glabrous. Heads in corymbs. Herbs.
  Fertile achenes compressed.
    Ray white or pink; fertile achenes very flat... *Cassinianum*
    Ray yellow; phyllaries obtuse, rugose ... *Ayersii*
  Fertile achenes not compressed.
    Ray pink or white; branches and peduncles long, slender ... ... *Lawrencella*
    Ray yellow or white; branches compact, peduncles very short ... ... *semifertile*
Flowers all fertile, some of the marginal ones pistillate only and without pappus
  Ray yellow or brownish; achenes glabrous. Heads solitary.
    Phyllaries wrinkled, obtuse. Woolly herbs.
      Stems branchless or shortly branched; heads large ... ... *scorpioides*
      Stems loosely branched; heads small; ray inconspicuous... ... ... *rutidolepis*
    Phyllaries smooth, obtuse. Glabrous or scantily hairy; leaves long, green, sometimes somewhat sticky ... ... *lucidum*
    Phyllaries narrow, acute, jagged; leaves stalked. Woolly. ... ... *podolepideum*
  Ray white or pinkish. Woolly herbs.
    Achenes papillary-rough; outer phyllaries woolly.
      Heads solitary, leaves small, rigid, linear, obtuse ... ... *obtusifolium*
      Heads corymbose; leaves lanceolate, soft, thick ... ... *Blandowskianum*
    Achenes glabrous. Heads solitary, terminal.
      Plant glandular-rough; leaves linear-revolute, upper ones clasping ... ... *adenophorum*

Plant woolly; leaves long, sessile, scabrous above, cottony-white below ... *leucopsidium*

Plant woolly; leaves narrow-linear, revolute; pappus-bristles somewhat *plumose* at the end; inner phyllaries *ciliate* ... *Baxteri*

  ii. Inner phyllaries without ray-like tips.
   a. Phyllaries yellow. Herbs.

Heads singly terminal; phyllaries in few rows (also *H. rutidolepis*).

 Stems or branches leafy, beset with glandular or cottony hairs; phyllaries shining, ciliate, acute; achenes scabrous ... *ambiguum*

 Leaves radical; leafy bracts clasping, phyllaries membranous; achenes glabrous. Minute annual, stems filiform ... *Tepperi*

Heads corymbose; phyllaries in several rows, ciliate; achenes glabrous; pappus-bristles plumose at the end.

 Leaves flat, cottony; heads loosely corymbose ... *apiculatum*

 Leaves linear, sticky; heads densely corymbose ... *semipapposum*

  b. Phyllaries white or pale-yellowish. Shrubs.

Heads rather large, solitary; branches spinescent, cottony ... *Dockerii*

Heads small in panicles; leaves lanceolate, long ... *Thomsoni*

Heads small in terminal corymbs.

 Leaves small, linear, more or less decurrent; achenes rough.

  Leaves strongly decurrent, truncate; phyllaries appressed ... *decurrens*

  Leaves faintly decurrent, retuse; phyllaries rather loose ... *retusum*

 Leaves not decurrent on the stem.

  Leaves flat, elongate, brownish below; phyllaries spreading, pale yellowish; achenes minutely hairy ... *ferrugineum*

  Leaves linear, margins revolute; phyllaries pale-yellow, appressed. Achenes papillary-rough. Coast shrub, somewhat sticky ... *cinereum*

   Achenes silky-hairy, surrounded at the base by a callous ring ... *Kempei*

Polycalymma.

Stout branchless herb, cottony or glandular-hairy. Head solitary, very large; partial heads 5- to 8-flowered, forming large corymbose-like clusters within the general involucre; leaves lanceolate, long ... *Stuartii*

Hyalolepis.

Dwarf, tufted; partial heads 1-flowered; pappus of one bristle or 0 ... *rhizocephala*

Decumbent; partial heads 4-flowered; pappus of 1 to 4 scales ... *Rudallii*

Angianthus.

Compound heads subtended by floral leaves.
   Spikes cylindrical; pappus of one oblique, fringed scale ... *pleuropappus*
   Spikes oblong; pappus very short, membranous ... *brachypappus*

Compound heads not subtended by floral leaves.
   Pappus of 2 or 3 scales, ending in bristles plumose at the end; spikes short-cylindrical; leaves linear to oblong-cuneate, cottony ... *tomentosus*

   Pappus absent or rudimentary. Annuals almost glabrous, the spikes yellow or brownish, lustrous.
     Spike oblong, attenuate at the base; pappus a minute ring ... *pussillus*
     Spike short, cylindrical, obtuse at both ends; pappus 0 ... *tenellus*

Skirrophorus.

Floral leaves linear, recurved-pointed; flowers solitary in each headlet; pappus absent ... *strictus*

Floral leaves from ovate- to narrow-lanceolate; flowers 2 in each headlet; pappus a minutely toothed ring ... *Preissianus*

Gnephosis.

Pappus of short plumose scales. Minute almost stemless plant ... *Burkitti*

Pappus absent; compound heads nearly globose.
   Stems prostrate; achenes and heads woolly ... *eriocarpa*
   Stems erect, branched; achenes glabrous ... *arachnoidea*

Pappus cup-shaped, slightly toothed or jagged; receptacle convex ... *cyathopappa*

Pappus cup-shaped, truncate; receptacle depressed ... *codonopappa*

Pappus tubular towards the base, lobed; receptacle convex ... *skirrophora*

Calocephalus.

I. Partial heads 2- or 3-flowered.

Leaves alternate, pappus plumose from the base. Dwarf annual, loosely woolly; leaves narrow-linear ... ... *Drummondii*
    Low rigid shrub, white with a close tomentum; leaves minute; compound heads globular, white. Sea-cliffs ... ... *Brownii*
Leaves alternate; pappus plumose at the end. Erect, cottony herb; compound heads ovoid or globose, yellow ... ... *Sonderi*
Leaves mostly opposite; pappus plumose at the end; erect slender herbs, with a greyish appressed investiture.
    Compound heads oblong-ovoid, white; leaves obtuse ... ... *lacteus*
    Compound heads yellow; leaves mostly acute ... *citreus*

II. Partial heads many-flowered.

Sub-shrubby, woolly-white; leaves linear; compound heads globose, finally irregularly lobed; pappus woolly-plumose ... ... *platycephalus*

Eriochlamys.

Diffuse woolly-tomentose annual; heads sessile amongst floral leaves; pappus absent; leaves short, linear *Behrii*
Larger stature; leaves lanceolate; pappus of a few plumose bristles ... ... *Knappii*

Cephalipterum.

Annual, stems single erect with a globular cluster of flower-heads ... ... *Drummondii*

Gnaphalodes.

Dwarf, diffuse, tomentose, annual; leaves stalked, obovate ... ... *uliginosum*

Pterocaulon.

Compound heads globular, solitary; leaves lanceolate *sphacelatus*
Heads in spicate clusters; leaves obovate or long ... *Billardieri*

Craspedia.

I. Compound heads singly terminating the stem.

Heads depressed-globular, pale yellow. Lower leaves large ovate- to narrow-lanceolate, stem-leaves clasping; outer phyllaries of general involucre ovate, conspicuous, with broad brown margins; pappus of 10 to 15 white plumose bristles ... ... *Richea*

Heads globular, yellow; outer phyllaries inconspicuous; pappus yellow.
    Leaves glabrous above; pappus-bristles plumose above the middle, connate at the base; heads ½ in. diameter ... *chrysantha*
    Leaves silky on both sides, bristles ciliate from the base, heads larger ... *globosa*
    II. Compound heads in clusters of 2 to 5, consisting of a large terminal head and 1 to 4 smaller and lateral ones.
Phyllaries of the general involucre conspicuous. Corolla bright-yellow; pappus-bristles yellow ... *pleiocephala*

Chthonocephalus.

Stemless; leaves oblong, flaccid, somewhat cottony ... *pseudevax*

Microseris.

Root fleshy, edible; leaves elongate, entire, or if pinnatifid the lobes short and narrow; corolla yellow ... *Forsteri*

ORDER CANDOLLEACEAE (Stylidieae).

Column elongate, bent down but suddenly becoming erect on being touched; the fifth corolla-lobe minute, narrow and immovable *(Stylidium)* ... **Candollea.**
Column erect, not elastic; the fifth corolla-lobe hooded, reflected but becoming erect when touched ... **Leewenhoekia**

Candollea.

    I. Leaves radical; capsule globular or ovoid.

Leaves linear or linear-lanceolate, very long. Tall, glabrous, except the glandular-hairy inflorescence; flowers pink, nearly sessile, racemose, on very long scapes; fruit ovoid-oblong ... *graminifolia*
Leaves narrow-linear, also at the summit of short simple branches; flowers rosy-red, few in a raceme. Dwarf, glabrous ... *Tepperiana*
Leaves narrow-elliptical to orbicular; flowers white, red-spotted, corymbose; fruit almost globular. Minute, glandular-hairy.
    Corolla with a spur-like prolongation ... *calcarata*
    Corolla without a spur ... *perpusilla*
Leaves spathulate; flowers panicled; fruit narrow-oblong ... *floribunda*

    II. Leaves scattered, linear; capsule linear.

Minute glabrous plant with white corymbose flowers ... *despecta*

Leewenhoekia.

Minute, erect, glandular annual; leaves ovate; flowers white, axillary and solitary forming a leafy raceme or corymb ... ... *dubia*

ORDER CAMPANULACEAE.

Corolla irregular, anthers connected, ovary 2-celled.
   Corolla-tube slit on one side to the base ... ... **Lobelia**
   Corolla-tube entire ... ... **Isotoma**
Corolla regular, anthers free, ovary 3-celled ... ... **Wahlenbergia**

Lobelia.

I. Flowers in terminal racemes; all the anthers hair-tufted.
Flowers singly terminating long branch-like stalks; lower leaves cuneate-obovate, incised; capsule bulging on upper side ... ... *rhombifolia*
Flowers in a one-sided raceme.
   Leaves linear, entire; stem erect, branchless, glabrous, turgid and somewhat fleshy; upper corolla-lobes glabrous; capsule very gibbous; seeds very minute ... ... *microsperma*
   Leaves linear, the lower ones broader and denticulated; stems slender, hardly succulent, glabrous; upper corolla-lobes minutely hairy.
     Capsule very gibbous; seeds very minute ... *Browniana*
     Capsule slightly gibbous; seeds small, ovate-triangular; leaves mostly indented ... ... *simplicicaulis*
   Leaves pinnatifid; stems simple or slightly branched, more or less hairy; seeds small, triangular, winged ... ... *heterophylla*

II. Flowers solitary, axillary; the two lower anthers hair-tufted. Perfect stamens and pistils in distinct flowers mostly on separate plants.
Flower-stalks elongate.
   Glabrous, procumbent or ascending; leaves ovate- to elliptical-lanceolate; 2 upper corolla-lobes short ... ... *purpurascens*
   Slightly pubescent, creeping; leaves ovate to orbicular; corolla-lobes nearly of equal size ... *pedunculata*
Flower-stalks shorter or not much longer than the leaves.
   Glabrous, erect; leaves ovate or oblong, serrate; corolla-lobes nearly equal, the two upper more deeply separate; fruit subglobular ... ... *concolor*

Glabrous, creeping; leaves linear-cuneate or oblong-spathulate; corolla-lobes nearly equal, oblique; fruit much compressed ... ... *platycalyx*

Glabrous, lax, sometimes creeping, prominently triangular stems; leaves cuneate or obovate; fruit cylindrical ... ... *anceps*

Pubescent, creeping; leaves linear to oblong, toothed; fruit dry ... ... *pratioides*

Pubescent, creeping; leaves ovate to orbicular, almost entire; fruit slightly succulent ... ... *Benthami*

Isotoma.

Erect and branching, growing in rock-fissures; flowers large, on long axillary stalks; leaves ovate or lanceolate, toothed ... ... *petraea*

Leaves radical, obovate or oblong; scapes slender erect *scapigera*

Creeping or prostrate; leaves ovate; flowers small, axillary ... ... *fluviatilis*

Wahlenbergia.

Stems leafy, simple or branched, erect, beset with short spreading hairs; flowers on long stalks, singly terminating stems or branches; leaves from ovate-lanceolate to linear ... ... *gracilis*

ORDER GOODENIACEAE.

I. Corolla-tube entire.

Anthers connate; fruit dry, indehiscent, *free from calyx* ... ... **Brunonia**

II. Corolla-tube slit on one side to the base.

Anthers connate, calyx-tube adnate to the ovary.
   Calyx lobed, fruit capsular ... ... **Leschenaultia**
   Calyx-lobes very small or absent; fruit dry, not bursting, 1-seeded; flowers blue ... ... **Dampiera**

Anthers free.
   Calyx free from the ovary; fruit capsular... ... **Velleya**
   Calyx-tube adnate with the ovary and fruit.
      Fruit more or less succulent, indehiscent.
         Ovules several, erect; succulent herb ... **Selliera**
         Ovules 2 in each cell, pendulous ... ... **Catosperma**
         Ovules 1 or 2 in the whole ovary; corolla-lobes expanding into broad glabrous wings as the flower opens; flowers blue. Shrubs or herbs ... ... **Scaevola**

Style undivided; corolla yellow ... ... **Goodenia**
Style 2-cleft ... ... **Calogyne**

Brunonia.

Flowers blue in a dense head on a scape from among basal leaves ... ... *australis*

Leschenaultia.

Leafless divaricate shrub, capsule beaked ... ... *divaricata*
Leaves filiform, branches striated, fruit not beaked ... *striata*

Dampiera.

Leaves flat, rigid, glabrous, oblong or linear; corolla beset with appressed brownish hairs; branchlets very angular ... ... *stricta*
Leaves flat, ovate, tomentose underneath; calyx beset with plumose hairs ... ... *candicans*
Leaves recurved at the margin.
   Corolla covered with a stellate tomentum; leaves elliptical to broad-linear, the margins much revolute, usually glabrous ... ... *rosmarinifolia*
   Corolla beset with a stellate tomentum; leaves narrow-elliptic to ovate; margins slightly revolute; cymes very short; flowers 1 to 3 on each stalk with leafy bracts ... ... *marifolia*
   Corolla with plumose hairs; leaves linear- to elliptic-lanceolate; flowers often in compound leafy racemes ... ... *lanceolata*

Velleya.

Bracts leafy, distinct; leaves radical, stalked, broadly ovate to narrow-lanceolate, often coarsely toothed; flowers in dichotomous cymes on long peduncles; corolla yellow with a hollow conical protuberance at the base ... ... *paradoxa*
Bracts connate, very large; sepals usually connate at the base ... ... *connata*

Selliera.

Glabrous prostrate perennial; leaves ovate to linear-spathulate; flowers on short stalks, solitary and axillary; corolla-lobes not winged ... ... *radicans*

Catosperma.

Glabrous, procumbent; leaves stalked, ovate; peduncles 3-flowered ... ... *Muelleri*

Scaevola.

1. Shrubs.

Flowers solitary, axillary, on slender peduncles; leaves entire.
    Leaves clustered, obovate to linear-elliptical, hoary; branchlets often spinescent; corolla bluish-white, its lobes somewhat fringed; drupe purple ... *spinescens*
    Leaves elliptic-ovate, scattered, glabrous; branchlets thornless... ... ... *Groeneri*
Flowers sessile, blue, in terminal spikes; bracts linear. Tall, viscid, glabrous; leaves large, thick, obovate-orbicular or spathulate, closely serrate ... ... *crassifolia*

II. Rigid herbs or almost herbaceous undershrubs.

Flowers axillary and solitary, or in irregular cymes in the lower axils.
    Stem-leaves mostly reduced to small bracts or wanting.
        Plant hirsute, erect and much branched ... *parvifolia*
        Plant glabrous with elongate divaricate branches... ... ... *depauperata*
    Stems leafy, glabrous; leaves linear-lanceolate; flowers yellowish, solitary, or clustered in the lower axils ... ... *collaris*
Flowers sessile or nearly so in the axils of floral leaves or bracts, the upper ones forming a terminal spike.
    Prostrate; leaves quite entire, somewhat fleshy; plant robust, beset with appressed hairs; flowers large in terminal spikes; throat of corolla villous with appressed hairs; fruit succulent ... *suaveolens*
    Procumbent; leaves rather small, cuneate-ovate, toothed. Flowers almost sessile; style and corolla-throat with short hairs; fruit very small, 1-seeded, dry *microcarpa*
    Diffuse, ascending, clothed with appressed hairs; leaves obovate or cuneate; indusium with a dense tuft of purplish hairs; bracteoles linear; ovary 2-celled
        Leaves coarsely toothed, bracteoles linear ... *aemula*
        Leaves acutely toothed; bracteoles ciliate ... *humilis*
    Erect, pubescent; leaves ovate or obovate, coarsely toothed; ovary 2-celled ... ... *ovalifolia*
    Erect, villous; leaves linear-revolute, entire; fruit 1-seeded; corolla pubescent outside, with short hairs in the throat ... ... *linearis*

Goodenia.

i. Flowers in panicles.

Panicle large, terminal; corolla blue ... ... *Ramelii*
Panicle small, from among crowded radical leaves; corolla yellow; seeds many, minute, orbicular, shining ... ... *humilis*

    ii. Flowers axillary, 1 to 3, with bracteoles. Stem leafy; corolla yellow. Shrubs.

Leaves clasping, serrate, glandular-viscid ... ... *amplexans*
Leaves stalked, orbicular to lanceolate; glabrous, viscid.
  Leaves rather thin, closely denticulated, stalked; fruit cylindrical, dissepiment reaching far above the middle. Moderately tall ... ... *ovata*
  Leaves rather thick, irregularly toothed, narrowed at the base; fruit rather short; dissepiment half as long as fruit. Dwarf ... ... *varia*

    iii. Flowers axillary, solitary; bracteoles absent or minute. Stems leafy. Corolla yellow in most species.

Stem-leaves all stalked; glandular-pubescent or glabrous, never villous.
  Leaves linear, corolla blue ... ... *Vilmoriniae*
  Leaves ovate, truncate or cordate at the base, toothed, on long stalks ... ... *grandiflora*
  Leaves orbicular, coarsely toothed, shortly stalked *Chambersii*
  Leaves ovate-lanceolate, narrowed at the base, acutely toothed; corolla white. Glabrous ... *albiflora*
  Leaves deeply lobed or pinnate.
    Corolla spurred; leaves deeply pinnatifid ... *calcarata*
    Corolla-tube protuberant, not spurred; leaves pinnate.
      Terminal leaflet cuneate at the base; lateral leaflets several. Tomentose undershrub ... ... *Nicholsoni*
      Terminal leaflet truncate or cordate; lateral leaflets few and small or wanting *grandiflora*

Stem-leaves abbreviated and sessile. Villous or silky hairy.
  Decumbent or ascending; leaves on long stalks, coarsely toothed; corolla laterally protuberant at the base.
    Densely villous-tomentose, corolla $\frac{3}{4}$ in. long... *Mitchellii*
    Softly pubescent, corolla under $\frac{1}{2}$ in. long ... *heterochila*
    Calyx-lobes *lanceolate, leafy;* hispid or viscid-villous ... ... *sepalosa*

  Erect, hoary-tomentose or silky; leaves remotely
   toothed ... ... *Mueckeana*
   Silky with appressed hairs; leaves entire,
    crowded ... ... *Strangfordii*

  IV. Flowers yellow, solitary on axillary scapes;
   leaves radical or tufted.

Bract present.
  Calyx-lobes linear, obtuse; leaves obovate to lanceolate, distantly toothed; flowers rather large on long scapes, which bear bracts at about the middle. Tufted, more or less hairy, rarely woolly, herbs... ... ... *geniculata*
  Calyx-lobes subulate; stems trailing; hispid ... *hirsuta*

Bracts not developed.
  Leaves softly villous or tomentose; seeds black, winged ... ... *cycloptera*
  Leaves glabrous or with scattered appressed silky hairs.

    a. Radical leaves entire.

   Radical leaves ovate to narrow-lanceolate; corolla silky outside; seeds broadly winged, dissepiment of fruit short ... ... *glauca*
    Dissepiment reaching above the middle of the fruit; calyx-teeth shorter; corolla-lobes narrower ... ... *microptera*
   Radical leaves ovate; stem-leaves few distant; corolla almost glabrous outside; seeds hardly winged; peduncles long and slender ... *elongata*
   Radical leaves broad- to linear-lanceolate; corolla with appressed hairs outside; dissepiment half as long as the fruit; seeds brown, about 20, with a narrow wing ... *heteromera*

    b. Radical leaves pinnatifid.

   Flowers large; seeds black with a broad wing; dissepiment short ... ... *pinnatifida*
   Flowers very small; calyx-lobes broader; dissepiment very short; indusium somewhat bilobed. Prostrate ... ... *pusilliflora*

Calogyne.

Erect glandular-hairy annual; dissepiment of fruit reaching to the middle; flowers yellow, stalked, axillary, ebracteate ... ... *Berardiana*

Sub-Class IV.—Choripetaleae Hypogynae.
ORDER PRIMULACEAE.

Calyx free, corolla rotate with a short inflated tube; capsule bursting transversely ... ... **Centunculus**

Calyx tubular; corolla salver-shaped, with scales between the lobes; capsule bursting by valves... ... ... **Samolus**

Centunculus.

Minute; leaves alternate, ovate; flowers axillary, solitary, nearly sessile, pale-rose, very small; stamens 4 ... *minimus*

Samolus.

Flowers white, rather large, in racemes; leaves firm, somewhat succulent, lanceolate or obovate; stamens 5 with 5 alternating staminodia. Perennial herb, erect and stoloniferous ... ... *repens*

ORDER CONVOLVULACEAE.

I. Leafy plants. Sepals distinct.

Style filiform undivided to the stigmatic lobes; ovary 2-celled, usually 2 ovules in each cell; flowers axillary.
Stigma-lobes 2, globular. Trailing or climbing herbs **Ipomoea**
Stigma-lobes 2, linear-oblong. Trailing or climbing herbs ... ... **Convolvulus**
Stigma-lobes 6 to 8; fruit 2-celled, 1 seed in each ... **Polymeria**

Style more or less branched below the stigmas. Ovary 2-celled, 2 ovules in each cell. Trailing or twining, subshrubby.
Style-branches 2, stigmas globular ... ... **Breweria**
Style-branches 4, stigmas linear ... ... **Evolvulus**

Styles 2. Small prostrate or diffuse perennials.
Stigmas globular; fruit 1-celled, 1-seeded ... ... **Cressa**
Stigmas globular, ovary of 2 carpels each with a separate style and ovule ... ... **Dichondra**

II. Leafy plants; calyx 5-toothed.

Style-branches 2, stigmas globular, fruit 1- to 2-seeded. Prostrate or diffuse, somewhat shrubby; flowers very small, axillary, solitary and sessile ... ... **Wilsonia**

III. Leafless parasites with filiform twining stems.
Styles 2, calyx lobed ... ... **Cuscuta**

Ipomoea.

Leaves of 3 obovate leaflets; stellately tomentose *Davenporti*

Leaves very obtuse, cordate, orbicular or reniform; glabrous ... *costata*
Leaves cordate-ovate, with basal lobes; glabrous ... *Muelleri*
Leaves lanceolate or oblong, entire or toothed; hairy ... *heterophylla*

Convolvulus.

Trailing perennial, glabrous or pubescent; leaves from cordate-sagittate to ovate-lanceolate, or the upper ones linear, lobed or entire; flowers pale-rose, solitary on longish stalks; bracts minute distant from the calyx; fruit 2-celled ... *erubescens*

Climbing perennial, glabrous; leaves large, broadly ovate-triangular to lanceolate-hastate; flowers solitary, white, on long stalks; bracts large enclosing the calyx; fruit 1-celled. River-banks ... *sepium*

Polymeria.

Stem slightly branched erect; leaves with appressed hairs; flowers pink about ¾ in. long ... *longifolia*
Prostrate, leaves densely silky-hairy; flowers smaller ... *angusta*

Breweria.

Tomentose undershrub, leaves obovate; bracteoles as long as calyx; flowers pink, solitary, about 1 inch ... *rosea*
Pubescent; lower leaves somewhat cordate, upper ones lanceolate; bracteoles minute; flowers white, smaller ... *media*

Evolvulus.

Prostrate or erect, leaves oblong or lanceolate, flowers blue 1 to 3 together on slender axillary peduncles ... *linifolius*

Cressa.

Erect or diffuse, much branched; flowers in terminal spikes ... *Cretica*

Dichondra.

Prostrate or creeping; leaves on long stalks orbicular or reniform; flowers solitary on short stalks ... *repens*

Wilsonia.

Branches prostrate, somewhat shrubby; leaves concave, imbricate, closely beset with grey shining appressed hairs ... *humilis*
Dwarf, herbaceous; leaves flat, thick, orbicular to ovate, not imbricate, with loose scattered hairs ... *rotundifolia*
Glabrous; leaves broad-linear, fleshy ... *Backhousii*

Cuscuta.

Flowers almost sessile in globular clusters, *glandular-dotted* — *australis*
Flowers distinctly stalked; corolla somewhat bell-shaped — *Tasmanica*

ORDER BORAGINEAE.

I. Ovary entire, style terminal; fruit dry.

Style bifid; fruit 4-lobed; leaves lobed ... ... **Coldenia**
Stigma hemispheric or conical with a fleshy ring round its base, corolla white; fruit 4-lobed ... ... **Heliotropium**
Style filiform, without an annular enlargement.
    Fruit separating into 2 fruitlets, each 2-celled, 2-seeded; anthers connate, terminating in long straight beaks ... ... **Halgania**
    Fruit separating into 4, 1-seeded, fruitlets; anthers cohering by their hairs, the points long and twisted ... ... **Pollichia**

    II. Ovary 4-lobed; style inserted between the lobes; fruit dry, separating into 2 or 4 nuts.

Nuts erect, laterally attached to the narrow conical receptacle.
    Nuts 2, wrinkled or granulate ... ... **Rochelia**
    Nuts 4, beset with hooked bristles ... ... **Echinospermum**
Nuts erect, obliquely attached, 4, usually reticulate-wrinkled ... ... **Eritrichium**
Nuts erect, fixed to the base only, 4, smooth and shining. Corolla with 5 small scales at the base of the lobes ... ... **Myosotis**
Nuts depressed, obliquely attached, 4, beset with hooked prickles ... ... **Cynoglossum**

Coldenia.

Prostrate hirsute annual, flowers sessile axillary ... *procumbens*

Heliotropium.

I. Stigma hemispheric, almost sessile; bracts absent.

Glabrous, somewhat succulent, prostrate; leaves from oval to almost linear; flowers small, white, sessile in a one-sided simple spike; fruit very short and broad ... ... *Curassavicum*
Hairy; leaves oblong to linear-lanceolate, crisped, dentate; fruitlets winged ... ... *pleiopterum*

    II. Stigma conical; anthers not cohering; ebracteate; throat of corolla not bearded.

Leaves ovate, flat, conspicuously stalked, beset with short appressed hairs; style shorter than the stigma; spikes once-forked ... ... *Europaeum*

Leaves oblong-lanceolate, waved, sessile or nearly so, rough with rigid hairs. Spikes once- or twice-forked.
    Style shorter than the long stigma; spikes elongated ... ... *undulatum*
    Style longer than the slender stigma; spikes short and dense ... ... *asperrimum*

    III. Stigma conical; anthers acuminate cohering by their minutely hairy tips; corolla-throat bearded.

Bracts absent. Leaves obovate-oblong or lanceolate, flat, white with appressed silky hairs; style shorter than stigma; spikes simple or once-forked ... ... *ovalifolium*

Bracts present; corolla-tube swollen round the anthers; style longer than the stigma.
    Bracts leafy; leaves oblong-lanceolate, woolly; flowers in dense terminal cymes ... ... *filaginoides*
    Bracts small; leaves narrow-linear, hoary; flowers in once- or twice-forked spikes ... *tenuifolium*

Halgania.

Leaves obovate- to linear-cuneate, with crisped margins and indented at the end; glandular glutinous; flowers deep-blue in cymous corymbs ... *cyanea*

Leaves lanceolar, thick, entire, with recurved margins, hoary-white below; flowers blue in small corymbs; anthers viscid ... ... *lavandulacea*

Pollichia.

Tall annual, flowers pale-blue in simple racemes, bracts leafy. ... ... *Zeylanica*

Rochelia.

Dwarf hispid annual; leaves linear; corolla minute, white ... ... *Maccoya*

Echinospermum.

Dwarf erect hairy annual, leaves ovate- to linear-elliptical, flowers small in leafy racemes; corolla blue ... ... *concavum*

Eritrichium.

Dwarf diffuse hairy annual, leaves linear; flowers small in the axils of leafy bracts, forming a one-sided raceme ... ... *Australasicum*

Myosotis.

Small erect or diffuse hispid annual; flowers very small, without bracts, in one-sided racemes ... *australis*

Cynoglossum.

Leaves lanceolate to oblong. Erect beset with rigid hairs.
    Fruitlets prickly all over.
        Flowers in bracteate racemes, pedicels longer than the calyx; corolla white ... *suaveolens*
        Flowers in leafless racemes, pedicels shorter than the calyx; corolla blue ... ... *australe*
    Fruitlets prickly on the raised margin only; ebracteate ... ... *Drummondii*

ORDER ASCLEPIADEAE.

Erect, with succulent jointed leafless branches.
    Pollen masses pendulous; corona inflated; fruitlets slender ... ... **Sarcostemma**
Leafy usually twining herbs or somewhat shrubby.
    Pollen masses pendulous; corona forming a loose cup, the margin 10-lobed, with the addition of 10 inner lobes; fruitlets more or less winged... ... ... **Cynanchum**
    Pollen masses pendulous; corona scarcely prominent; fruitlets slender ... ... **Daemia**
    Pollen masses erect; corona compressed; fruitlets large, thick, ovoid ... ... **Marsdenia**

Sarcostemma.

Flowers small in lateral umbels. About 1 to 2 ft. high... *australe*

Cynanchum.

Leaves cordate on long stalks; flowers in dense cymes; corona-lobes subulate; inflorescence pubscent ... *floribundum*

Daemia.

Leaves linear, glabrous; umbels few-flowered, corolla green-purplish ... ... *Kempeana*

Marsdenia.

Leaves 2 to 4 in. long; flowers in simple, dense umbels... *Leichhardtiana*

ORDER APOCYNEAE.

Carpels united; ovary 2-celled with axillary placentas; fruit a drupe. Erect shrub, more or less thorny ... **Carissa**

Carpels 2, distinct, ovules few in each; fruitlets drupaceous. Shrub ... **Alyxia**

Carpels 4 or 2-bipartite, 1-ovulate ... **Notonerium**

Carissa.

Leaves from orbicular to lanceolate, glabrous or young leaves pubescent; flowers in sessile or shortly stalked axillary cymes ... *Brownii*

Alyxia.

Bushy shrub, glabrous; leaves orbicular-to elliptic-ovate; flowers in terminal sessile heads or clusters; fruitlets orange ... *buxifolia*

Notonerium.

Subshrubby; leaves alternate linear; flowers small in terminal cymes ... *Gossei*

ORDER GENTIANEAE.

Leaves opposite; corolla-lobes without any expanding membranes. Terrestrial herbs, rather small, erect, glabrous.
  Corolla-tube cylindrical; style deciduous.
    Calyx divided nearly to the base; anthers at length straight; fruit 2-celled; flowers yellow, in cymes; leaves sessile ... **Sebaea**
    Calyx shortly lobed; anthers at length twisted; fruit 1-celled; corolla red ... **Erythraea**
  Corolla-tube campanulate; style persistent; fruit 1-celled. Corolla bluish or white ... **Gentiana**

Leaves radical or scattered; corolla-lobes with expanding membranes. Floating or semiaquatic; corolla yellow ... **Limnanthemum**

Sebaea.

Corolla bright-yellow, 5-lobed; calyx-segments acute ... *ovata*

Corolla yellowish-white, 4-lobed; calyx-segments obtuse ... *albidiflora*

Erythraea.

Flowers nearly sessile, often forming somewhat one-sided and cymous spikes; leaves from oval- to narrow-elliptical ... *spicata*

Gentiana.

Rather tall; leaves stalked obovate or spathulate; flowers large on long stalks often forming a compact corymb ... *saxosa*

Limnanthemum.

Leaves crenate, floating, on long stalks, cordate-orbicular; flowers clustered on long pedicels ... ... *crenatum*

Erect, semiaquatic; leaves ovate- to orbicular-cordate on long stalks; flowers in a cymous panicle... ... ... *reniformis*

ORDER JASMINEAE.
Jasminum.

Corolla-lobes 5 or more. Ovules ascending or erect. Fruit succulent.
　Leaves of 3 narrow long leaflets, calyx-lobes exceedingly short; flowers white in racemose panicles; berries black. Shrubby, climbing to many feet ... ... *lineare*

　Leaves simple; calyx-teeth equal or shorter than the tube *calcareum*

ORDER PLANTAGINEAE.
Plantago.

Leaves lanceolate, entire or short-lobed, hairy; flowers somewhat scattered, sessile, on elongated scapes; sepals obtuse, with broad scarious margins ... ... *varia*

ORDER LOGANIACEAE.

Corolla-lobes 4, valvate; calyx 2- to 4-lobed; styles two, nearly always connate at the summit; herbs ... **Mitrasacme**

Corolla-lobes 5, imbricate; calyx 5-cleft; style one. Shrubs ... ... **Logania**

Mitrasacme.

Prostrate hairy perennial, somewhat shrubby; calyx-lobes 4, ovate ... ... *pilosa*

Minute, erect, glabrous annuals; calyx 2-lobed.
　Styles cohering at the summit; leaves linear to lanceolate; flowers in terminal umbels or only 1, 2 or 3 together ... ... *paradoxa*

　Styles distinct; flowers often solitary and terminal, or 2 or 3 together and axillary ... ... *distylis*

Logania.

I. Leaves well developed; calyx-segments blunt; stamens inserted low down in the corolla-tube, anthers included

Leaves lanceolate-acuminate; flowers in pedunculate cymes leafy at the base. An erect glabrous tall shrub ... *longifolia*

Leaves obovate to orbicular, very thick, smooth and shining; flowers sessile in compact cymes... ... ... *crassifolia*

Leaves from broadly cordate-ovate to oval-elliptical ... *ovata*

Leaves linear with revolute margins ... ... *stenophylla*

Leaves broadly linear, flat; flowers small in compact stalked cymes; corolla nearly glabrous inside ... *linifolia*

III. Leaves minute or wanting; calyx-segments acute.

Stamens inserted in the throat of the corolla and exsert, flowers clustered at the nodes ... ... *nuda*

ORDER SOLANACEAE.

I. Stamens 5. Fruit a berry.

Corolla with a short wide tube. Shrubs or herbs ... **Solanum**
Corolla with a long narrow tube. Shrub ... ... **Lycium**

II. Stamens 5. Fruit dry, bursting in valves.

Corolla small broadly bell-shaped; *anthers 1-celled* ... **Anthotroche**
Corolla with a long narrow tube; flowers white.
    Calyx tubular, separating near the base after flowering; fruit prickly, reflexed ... ... **Datura**
    Calyx bell-shaped, persistent; fruit smooth ... ... **Nicotiana**

III. Stamens 4, very unequal.

Fruit a berry. Corolla-lobes short, obtuse. Tree ... **Duboisia**
Fruit capsular. Corolla-lobes long, spreading. Shrubs **Anthocercis**

Solanum.

I. Plant glabrous, without prickles.

Annual; leaves ovate on long stalks; flowers small, white, in umbels; berry globular, black ... ... *nigrum*

Shrubs; flowers large blue in lateral racemes.
    Leaves lanceolate, acute, the longer ones with a few lobes; berries yellow ... ... *aviculare*
    Leaves lanceolate, obtuse; berries purple.
        Berry globular ... ... *simile*
        Berry ovoid ... ... *fasciculatum*

II. Fruit stellately pubescent; no prickles on calyx; corolla blue.

Leaves glabrous above, velvety below; flowers small; berry red ... ... *ferocissimum*

Leaves velvety on both sides.
    Leaves orbicular, shortly stalked; corolla deeply lobed   *orbiculatum*
    Leaves cordate, almost sessile; corolla-lobes short ...   *oligacanthum*
    Leaves more or less lanceolate, conspicuously stalked.
        Corolla deeply lobed, leaves entire or shortly lobed towards the base; berry yellow. Dwarf, erect ... ... *esuriale*
        Corolla-lobes short and broad. Somewhat tall.
            Leaves mostly lobed or hastate; corolla $\frac{1}{2}$ in. diameter ... ... *chenopodium*
            Leaves entire or scarcely sinuate; corolla larger; berry black ... ... *Sturtianum*

    II. Prickles on the calyces, branches and leaves; plant usually stellately hairy; corolla blue.

Leaves glabrous, pinnatifid; corolla prickly. Diffuse, rigid ... ... *hystrix*

Leaves sprinkled with stellate hairs, scarcely tomentose, green on both sides, sinuate-lobed. Racemes short, few-flowered ... ... *eremophilum*

Leaves glabrous above, tomentose below, pinnatifid; calyx-lobes short, broad; corolla with deep acute lobes   *lacunarium*

Leaves tomentose on both sides; corolla-lobes short, broad.
    Leaves sinuate-lobed; calyx-lobes narrow, acuminate   *petrophilum*
    Leaves entire or slightly sinuate; calyx-lobes broad   *ellipticum*

Lycium.

Intricate shrub, branches spinescent; leaves clustered fleshy; flowers small, white, solitary; berry ovoid, red ... ... *australe*

Anthotroche.

Leaves orbicular, tomentose; flowers almost sessile; corolla hoary ... ... *Blackii*

Datura.

Erect, glabrous, annual; corolla large white; leaves large, lobed ... ... *Leichhardtii*

Nicotiana.

Erect, herb, beset with soft viscid hairs; flowers in loose terminal racemes, corolla white or greenish; lower leaves ovate on long stalks ... ... *suaveolens*

K

Duboisia.

Leaves linear, glabrous; corolla-lobes obtuse, flowers paniculate ... *Hopwoodii*

Anthocercis.

I. Leaves beset with glandular hairs.

Branchlets spinescent; leaves cuneate. Erect, intricately branched ... *anisantha*

Tall, branches twiggy; leaves narrow-linear; corolla yellowish-white; flowers terminal, large ... *angustifolia*

Diffuse, dwarf; leaves oval, small; flowers very small, solitary ... *myosotidea*

II. Leaves glabrous; branchlets and young foliage minutely hairy.

Erect; leaves large, elliptical; flowers whitish in leafy panicles ... *Eadesii*

ORDER EPACRIDEAE.

I. Ovules solitary in each cell; ovary usually 5-celled; fruit indehiscent, a dry or succulent drupe.

Corolla-lobes imbricate in the bud; throat of corolla closed with reflexed hair or scales ... **Brachyloma**

Corolla-lobes valvate in the bud ... **Styphelia**

II. Ovules several in each cell; fruit capsular.

Leaves stalked; corolla-lobes imbricate ... **Epacris**

Leaves clasping; corolla-lobes valvate ... **Sprengelia**

Brachyloma.

Corolla-lobes obtuse; calyx and corolla reddish; bracts several; leaves small, broad-linear to linear-lanceolate ... *ericoides*

Corolla-lobes acute, corolla whitish, bracts 2; leaves oval-elliptical or oblong-lanceolate.

Leaves almost blunt; corolla-lobes hardly imbricate *daphnoides*

Leaves flat, very small, shortly pungent-pointed ... *ciliatum*

Styphelia.

I. Anthers exserted; filaments filiform.

Corolla-tube with 5 dense tufts of hairs below the middle, corolla yellowish or pale reddish. Prostrate ... *adscendens*

Corolla-tube slender, glabrous inside ... *pusilliflora*

II. Anthers included; filaments flat; corolla-lobes bearded or hair-tufted inside; flowers red (*Astroloma*).

Corolla with 5 fringed scales inside near the base, not bearded at the throat.   Erect; flowers large, sessile   *Sonderi*

Corolla-tube with 5 tufts of hair inside, below the middle; and more or less hairy at the throat.   Prostrate   ...   *humifusa*

III. Anthers included; filaments filiform or terete.

A. *Corolla-lobes glabrous inside (Lissanthe).*

Corolla-tube cylindrical, more or less hairy above the middle; flowers yellowish; leaves doubly grooved below; pungent. Erect...   ...   .   *strigosa*

B. Corolla-lobes densely bearded, the tube glabrous or hairy inside above the middle; flowers white *(Leucopogon).*

a. *Flowers in terminal spikes; styles short.*

Tall shrubs, ovary 4- to 5-celled; spikes cylindrical.
  Leaves recurved at the margin, narrow, lanceolate, long; fruit depressed-globular, yellowish, succulent   *australis*
  Leaves flat, shorter and broader; fruit ovate-globular, whitish and succulent   ...   .   *Richei*

Undershrubs; ovary 2- rarely 5-celled; spikes short.
  Leaves *strongly 5-nerved*, cordate- to lanceolate-ovate, concave.
    Leaves thick, rather blunt, the lateral nerves divergent; flowers small, crowded   ...   ...   *costata*
    Leaves straighter, flatter, abruptly enlarged at the base; veins less-curved; sepals blunter ...   *striata*
  Leaves oblong or linear.
    Margins recurved or revolute.   Erect   ...   ...   *collina*
    Prostrate, *hirsute*   ...   .   *hirsuta*
  Leaves broadly lanceolate, concave, *distant* ...   ...   *concurva*
  Leaves lanceolate, *ciliate*, concave; *ovary 5-celled*   ...   *virgata*

b. *Flowers axillary, spicate or few-clustered. Style slender and long.*

Leaves oblong-linear, *margins recurved;* flowers few together; ovary 5-celled, hairy   ...   .   *ericoides*

Leaves *flat* or slightly concave, nerves very fine; ovary 5-celled.
  Corolla-tube shorter than the calyx; leaves ovate- to orbicular-cordate, very spreading   ...   .   *cordifolia*
  Corolla-tube scarcely exceeding the calyx; leaves oblong-elliptical, *hairy*, spreading ...   .   *hirtella*

Leaves concave or keeled; peduncles 1- to 3-flowered.
    Leaves cordate-ovate, pungent-pointed, minutely denticulated; flowers erect ... ... *rufa*
    Leaves ovate, appressed; flowers on recurved stalks *Woodsii*
        C. Corolla-lobes with a tuft of long hairs at the tips; throat hairy *(Acrotriche)*.
Flower-spikes or -clusters mostly in the axils of the previous year's leaves.
    Leaves linear-lanceolate, pungent; corolla green ... *serrulata*
    Leaves ovate-lanceolate, pungent; ovary 5-celled ... *patula*
    Leaves broadly ovate, obtuse; ovary 4-celled; corolla white inside ... ... *ovalifolia*
Flower-clusters below the leaves on the old branches.
    Leaves ovate to lanceolate, crowded, short; calyx narrow, reddish ... ... *depressa*
    Leaves lanceolate or linear-lanceolate, about ½ inch; calyx narrow, red ... ... *fasciculiflora*
        D. Corolla-lobes glabrous, tube companulate; ovule one *(Monotoca)*.
Small tree; flowers white, short-stalked, in short axillary racemes; leaves rather large, ovate ... ... *elliptica*

Epacris.

Corolla-tube much longer than the calyx.
    Rather tall, almost branchless; flowers solitary, axillary, forming one-sided spikes; corolla red, rarely white, with 5 impressions near the base; leaves small, lanceolate ... ... *impressa*
Corolla-tube shorter than the calyx, corolla white.
    Leaves very blunt; bracts and sepals blunt ... ... *obtusifolia*
    Leaves pungent-pointed.
        Leaves linear or lanceolate; bracts and sepals acute, fringed with minute hairs ... ... *lanuginosa*
        Leaves broad and cordate, bracts and sepals blunt *microphylla*

Sprengelia.

Erect; corolla pink, deeply cleft, about as long as calyx *incarnata*

ORDER LABIATAE.

Calyx with 5 nearly equal teeth.
    Stamens 4, equal; fruitlets smooth. Herbs... ... **Mentha**
    Stamens 4, in pairs, anthers 1-celled; fruitlets rugose.
        Upper lip of corolla widely separated from the lower; stamens exsert. Herbs or subshrubby **Teucrium**

GENERA AND SPECIES.

Upper lip of corolla very short; stamens hardly exsert ...	**Ajuga**
Stamens 4, in pairs, the lower pair sterile. Fruitlets rugose. Undershrubs. Upper lip of corolla hood-shaped ...	**Microcorys**
Upper lip of corolla flat, 2-lobed; leaves rigid, whorled ...	**Westringia**
Stamens 4, in pairs, the upper reduced to staminodia; calyx-teeth rarely 4; fruitlets smooth ...	**Lycopus**
Calyx-teeth unequal; stamens 4, in pairs; fruitlets smooth. Upper calyx-lobe broad, the lower ones narrow-pointed; flowers in leafless racemes; anthers 1-celled ...	**Plectranthus**
Upper calyx-lip with 3 teeth, the lower with 2; flowers in bracteate spikes; anthers 2-celled ...	**Prunella**
Calyx bilobed, the lobes entire; stamens 4, in pairs. Upper calyx-lobe with a hollow protuberance. Upper stamens 1-celled; fruitlets granulate. Herb ...	**Scutellaria**
Anthers 2-celled, appendaged; fruitlets rugose. Shrubs ...	**Prostanthera**

Mentha.

Leaves entire or scantily toothed; flowers in dense axillary clusters on very short stalks, forming terminal spikes.	
Leaves lanceolate; calyx-teeth subulate; corolla white. Erect. ...	*australis*
Leaves small, ovate; calyx-teeth short, lanceolar; corolla lilac. Dwarf ...	*gracilis*
Leaves very small; calyx-teeth short, densely hairy inside; corolla blue. Dwarf ...	*satureioides*

Teucrium.

Flowers sessile in leafy spikes; leaves 3- to 5-lobed. Dwarf herb ...	*sessiliflorum*
Flowers stalked in racemes or panicles; corolla white. Plant green; leaves entire; peduncles slender, 1-flowered ...	*integrifolium*
Plant green; leaves indented; peduncles slender, 3- or more-flowered ...	*corymbosum*
Plant hoary; leaves entire; peduncles rigid, 1-flowered ...	*racemosum*

Ajuga.

Erect herb, leaves chiefly radical; flowers blue, nearly sessile ... *australis*

Microcorys.

Glabrous; leaves in whorls of three, linear-terete, acute ... *Macrediana*

Westringia.

Leaves very much revolute, rigid, mostly in three's; calyx-teeth very short; flowers axillary, nearly sessile; corolla hirsute ... *rigida*

Leaves longer, less rigid, mostly in four's ... *Dampieri*

Lycopus.

Tall, erect, glabrous; leaves long, lanceolate; corolla white ... *australis*

Plectranthus.

Flowers small, blue, in whorls of 10; leaves on long stalks ... *parviflorus*

Prunella.

Procumbent herb, flowers purplish in terminal spikes ... *vulgaris*

Scutellaria.

Rather dwarf, slender, glabrous; corolla blue; flowers solitary ... *humilis*

Prostanthera.

1. Corolla-tube short and broad; upper lip very broad, short; the lower lip much longer with a large spreading middle lobe. Calyx-lips closed over the fruit.

Flowers in terminal leafless racemes.

Leaves oblong-lanceolate, glabrous; corolla white, purple-dotted; beset with minute hairs. Tall shrub or small tree ... *lasiantha*

Leaves ovate-orbicular, thick, slightly hoary; corolla lilac. Tall ... *rotundifolia*

Flowers axillary, or the upper ones crowded in leafy racemes.

Leaves oblong-lanceolate; corolla white, streaked with blue; upper calyx-lip ovate ... *striatiflora*

Leaves lanceolate; corolla white; upper calyx-lip cordate-deltoid ... *Wilkieana*

Leaves ovate to oblong-linear, small; calyx-lips nearly equal ... *eurybioides*

Flowers all axillary.
    Leaves ovate, glabrous; branchlets spinescent; corolla lilac ... ... *spinosa*
    Leaves oblong-linear; branchlets hoary; corolla white. Tall ... ... *Behriana*
    Leaves linear-terete, channelled above; branchlets tomentose ... ... *Baxteri*

        II. Corolla-tube incurved dilated upwards; upper lip concave, longer than the lower. Calyx-lips not closed over the fruit.

Leaves on rather long stalks, orbicular or broadly ovate   *ringens*

Leaves scarcely stalked, small, margins recurved.
    Corolla red; flower-stalks long... ... ... *coccinea*
    Corolla greenish; flower-stalks longer than calyx ... *chlorantha*
        Flower-stalks very short; corolla small, calyx large with broad and deep lobes ... ... *calycina*

ORDER LENTIBULARINEAE.

Segments of calyx two; aquatic or marsh plants ... **Utricularia**

Segments of calyx four in pairs, the inner ones lateral; no capillary ramifications ... ... **Polypompholyx**

Utricularia.

Stems floating, branched; submerged leaves divided into capillary segments beset with minute bladders; flowers yellow, in racemes on axillary stalks ... ... *flexuosa*

Stems branchless; leaves radical, minute; capillary ramifications very limited, bladders few. Marsh plants.
    Flowers purple, terminal in 1- or few-flowered scapes ... *dichotoma*
    Flowers purplish, almost sessile, distant along the scapes *lateriflora*

Polypompholyx.

Flowers small, pink; leaves radical, elliptical. Dwarf marsh plant ... ... *tenella*

ORDER OROBANCHEAE.

Orobanche.

Stems simple, robust about 1 foot; flowers purplish-blue in racemes ... ... *Australiana*

ORDER SCROPHULARINEAE.

I. Calyx lobed or toothed.

Calyx 5-lobed or -toothed. Stamens 4.
  Stems more or less leafy.
    Leaves constantly opposite; calyx toothed, tubular ... ... **Mimulus**
      Lower leaves opposite, upper ones scattered; calyx lobed.
        Stigma bilobed; leaves chiefly radical; flowers stalked ... ... **Mazus**
        Stigma entire; flowers sessile. Erect herb ... ... **Buechnera**
    Leaves stalked in basal tufts; calyx toothed, stigma globular. Aquatic or semi-aquatic ... ... **Limosella**
Calyx 5-toothed; stamens 2; corolla 5-lobed. Prostrate herbs ... ... **Peplidium**
Calyx 3- or 4-lobed; stamens 4; stigma spathulate; flowers solitary. Small creeping herbs ... ... **Glossostigma**
Calyx 4-lobed, stamens 4; stigma almost entire; corolla tubular, 2-lipped; flowers in leafy spikes. Erect herbs ... ... **Euphrasia**

II. Calyx divided to the base or nearly so.

Stamens 4; calyx-segments 5; corolla tubular 5-lobed; stigma bilobed. Erect herbs, somewhat shrubby... **Stemodia**
Stamens 2, staminodia 2 or 0; calyx-segments 5; corolla tubular at the base, 2-lipped, pale-pink; stigma dilated; flowers solitary, axillary. Erect herbs ... ... **Gratiola**
Stamens 2; calyx-segments 4 or 5; corolla rotate, or the tube much shorter than the lobes; stigma capitate ... ... **Veronica**

Mimulus.

Small, erect, glabrous; leaves lanceolate; flowers on long stalks; corolla blue... ... ... *gracilis*
Stems prostrate or creeping.
  Glabrous; leaves thick, ovate or oblong; flowers on short stalks, axillary, solitary; corolla lilac, throat yellow ... ... *repens*
  Pubescent; leaves very small, narrow-oblong; flowers smaller on rather long stalks; corolla-tube long and slender ... ... *prostratus*

Mazus.

Dwarf perennial; leaves obovate, irregularly toothed; flowers blue, in a terminal one-sided raceme or solitary, on long peduncles ... ... *pumilio*

Buechnera.

Scabrous-pubescent, upper leaves linear, flowers in slender spikes ... ... *linearis*

Limosella.

Flowers stalked; leaves linear to oblong, small; corolla and capsule exceeding the calyx ... ... *aquatica*

Flowers sessile, larger; leaves large, ovate; corolla and capsule included... ... ... *Curdieana*

Peplidium.

Flowers axillary, sessile or nearly so; capsule globular, obtuse ... ... *humifusum*

Flowers conspicuously stalked; capsule ovoid, acute ... *Muelleri*

Glossostigma.

Flower-stalks exceeding the minute linear-spathulate or oblong leaves; calyx usually 3-lobed; stamens as long as corolla ... ... *Drummondii*

Flower-stalks shorter than leaves; calyx 4-lobed; stamens shorter than the ovate *fringed* corolla-lobes ... ... *elatinoides*

Euphrasia

Flowers white; hairy perennial; leaves sessile, deeply serrate... ... ... *Brownii*

Flowers yellow; glandular-hairy annual; leaves sessile ... *scabra*

Stemodia.

Placentas consolidated into a single column.
    Glabrous; leaves linear or lanceolate; flowers solitary, sessile ... ... *Morgania*

Placentas free; flowers stalked; glandular-pubescent.
    Leaves lanceolate, serrate, sessile or stem-clasping, opposite or in whorls of three ... ... *viscosa*
    Leaves rhomboid-lanceolate, coarsely toothed, conspicuously stalked ... ... *pedicellaris*

Gratiola.

Flowers on long stalks; leaves lanceolate, glandular-hairy *pedunculata*

Flowers nearly sessile; leaves orbicular to ovate-lanceolate *Peruviana*

Veronica.
I. Shrubs.
Racemes short in terminal leafy panicles; leaves linear ... *decorosa*
Racemes elongate; leaves broadly lanceolate, serrate ... *Derwentia*
II. Perennial herbs, flowers in axillary racemes.
Leaves lanceolate, mostly sessile, nearly or quite entire ... *gracilis*
Leaves ovate, almost sessile, distantly toothed; flowers large; calyx small ... *distans*
Leaves broadly ovate, somewhat cordate, stalked, toothed; flowers smaller; calyx large ... *calycina*
III. Annual herb; flowers in leafy spikes.
Lowest leaves ovate, stalked, entire or serrate; flowers small ... *peregrina*

ORDER BIGNONIACEAE.
Tecoma.
Woody climber; stamens included, leaves pinnate, seeds winged. Flowers yellowish-white in loose panicles; calyx small, toothed ... *australis*

ORDER ACANTHACEAE.
Corolla 2-lipped, stamens 2. ... **Justicia**
Corolla-lobes nearly equal; stamens 4 ... **Ruellia**

Justicia.
Flowers pink in dense terminal bracteate spikes, or 1 or 2 pairs axillary.
   Erect, shrubby annual; leaves oblong-lanceolate or linear ... *procumbens*
   Dwarf, shrubby with spreading spinescent branchlets ... *Bonneyana*
Flowers axillary, solitary; leaves orbicular; slender pubescent herb ... *Kempeana*

Ruellia.
Bracteoles linear-subulate, shorter than calyx; flowers blue, axillary, sessile. Erect or diffuse ... *australis*
Bracteoles broad and long; corolla-tube shortly slender at the base ... *primulacea*

ORDER PEDALINEAE.
Josephinia.
Erect or diffuse, villous, herb; flowers pink, very small, axillary ... *Eugeniae*

ORDER VERBENACEAE.

i. Corolla 5-lobed; fruit 4-celled.

Corolla-lobes nearly equal.
    Fruit dry, separating into 4 fruitlets; stamens 4 or 2 ... ... **Verbena**
    Fruit dry, not separating into fruitlets. Woolly fructicose shrubs.
        Style undivided; stamens 5 ... ... **Newcastlia**
        Style deeply divided; stamens 5 ... ... **Dicrastylis**
    Fruit drupaceous; stamens 4; style 2-lobed ... **Clerodendrum**
Corolla 2-lipped or unequally lobed.
    Fruit drupaceous; stamens 4; style deeply cleft **Spartothamnus**

ii. Corolla 4-lobed, fruit opening in 2 valves, seed solitary.

Corolla-lobes nearly equal, stamens 4. Maritime shrub ... ... **Avicennia**

Verbena.

Spikes long and slender; lower leaves coarsely toothed, stalked; upper ones deeply divided; corolla not 2 lines long ... ... *officinalis*
Spikes short, very *glandular-hirsute;* flowers larger ... *macrostachya*

Newcastlia.

Flowers in terminal spikes.
    Corolla-lobes short, stamens included. Beset with loose woolly hairs ... ... *cladotricha*
    Corolla-lobes pointed; stamens exsert. Beset with a close tomentum... ... ... *spodiotricha*
Flowers in terminal heads; corolla-throat bearded ... *cephalantha*
Flowers axillary; corolla-lobes narrow.
    Stamens exsert; bracts membranous, imbricate, cordate ... ... *bracteosa*
    Stamens included; corolla slightly bearded inside near the base ... ... *Dixoni*

Dicrastylis.

i. Leaves stalked, flat.

Flowers in a pyramidal panicle; leaves lanceolate, rugose *ochrotricha*
Flowers in heads; leaves ovate or oblong-lanceolate; sepals subulate ... ... *Gilesii*

ii. Leaves sessile, margins recurved.

Leaves ovate or oblong-lanceolate; flower-heads on long stalks forming panicles ... ... *Doranii*

Leaves broadly linear; flowers in leafy slightly branched panicles ... *Beveridgei*

Flowers in small heads terminating the branches ... *Lewellini*

Clerodendrum.

Tall shrub; leaves ovate, stalked; flowers in cymes; stamens exsert ... *floribundum*

Spartothamnus.

Silky undershrub, leaves minute; corolla white, silky outside ... *teucriiflorus*

Avicennia.

Leaves coriaceous, ovate-lanceolate, closely tomentose below ... *officinalis*

ORDER MYOPORINEAE.

Corolla short, campanulate, nearly regular, white; ovary 2- or 4-celled, 1 ovule in each cell. Small trees or shrubs ... **Myoporum**

Corolla long, tubular, irregular, variously coloured; ovary 2-celled, usually 2 or more ovules in each cell. Shrubs, rarely trees ... **Eremophila**

Myoporum.

1. Fruit globular or nearly so; usually 3- or more-celled.

Perfect stamens 4; corolla bearded inside; leaves flat.
Corolla-lobes shorter than the tube; leaves narrow-lanceolate, acute, on long stalks. A very tall shrub *montanum*

Corolla-lobes as long as the tube.
Leaves thick, obovate-oblong, obtuse, bluntly toothed; fruit globular, succulent, black. Small tree or shrub ... *insulare*

Leaves thin, oblong or lanceolate, closely serrate; fruit ovate-globular, rather dry; shrub with viscid branchlets ... *viscosum*

Perfect stamens 5; corolla glabrous inside; leaves lanceolate, entire, flat; fruit 2- or 3-celled, somewhat succulent. Shrub ... *deserti*

Perfect stamens 4; corolla glabrous inside; leaves linear, short, thick. Procumbent or diffuse shrubs.
Corolla-lobes as long as the tube; peduncles comparatively long, 1 to 3 together; fruit nearly globular ... *humile*

Corolla-lobes shorter than the tube; peduncles short; fruit ovoid ... *brevipes*

GENERA AND SPECIES.

II. Fruit compressed, 2-celled.

Leaves linear-lanceolate, acute, serrate towards the summit; fruit small, dry, almost ovate, much flattened. Tree ... ... *platycarpum*

Eremophila.

1. Corolla-lobes not very unequal.

a. Calyx-segments imbricate at the base, remaining foliaceous.

Stamens included shorter than the corolla.
Leaves opposite, hoary; corolla campanulate, much narrowed and long at the base.
 Leaves narrow-linear, about 1 inch long, slender *Dalyana*
 Leaves narrow-linear, about ½ in, *recurved-pointed;* calyx very small; corolla blue, the lobes very short ... ... *scoparia*
 Leaves obovate about ¼ inch long... ... ... *Delisserii*
Leaves alternate, glabrous, not linear.
 Leaves very thick, complicated, recurved-pointed *crassifolia*
 Leaves obovate, toothed; sepals narrow-lanceolate; flowers bluish, axillary, solitary. Erect shrub ... ... *Behriana*
 Leaves cuneate-lanceolate, entire; sepals small, subulate ... ... *Weldii*
 Leaves oblong-lanceolate; flowers sessile; sepals linear ... ... *Christophori*
Leaves alternate, narrow, sessile.
 Sepals equal, narrow; leaves small.
  Leaves linear, acute, crowded; sepals narrow acute ... ... *densifolia*
  Leaves linear-oblong, *resinous-warty;* sepals subulate ... ... *gibbosifolia*
  Branchlets spinescent; sepals 4, lanceolate; leaves linear; corolla blue, hairy outside *divaricata*
 Sepals unequal, the 2 lower ones broader; leaves large.
  Leaves narrow-linear; calyx-segments ovate, with spreading or recurved points; corolla yellow; fruit tapering ... ... *polyclada*
  Leaves linear, calyx-segments broad-lanceolate, drupe hairy ... ... *Goodwinii*
  Leaves lanceolate, entire; drupe glabrous ... *Elderi*
  Leaves ovoid-elliptic, *deeply-serrate;* ovary tomentose ... ... *Willsii*

Leaves alternate, narrow-lanceolate, attenuated at
the base.
 Sepals equal; ovary glabrous ... ... *santalina*
 Sepals unequal, small, very acute, woolly at the
 margins; corolla dull-red; leaves linear-lanceolate, about 4 in. ... ... *longifolia*
  Calyx-segments lanceolate and ovate, corolla
  lavender; leaves lanceolate, acute, about
  1 in. ... ... *Freelingii*
Stamens, 2 or all, exceeding the corolla (also *E. longifolia*).
 Leaves lanceolate; fruit dry, ovate; calyx-segments
 unequal ... ... *bignoniflora*

  b. Calyx-segments not overlapping at the base.

Calyx campanulate, 5-lobed; flowers solitary, blue;
stamens included; leaves oblong-linear. Dwarf shrub *MacDonnellii*
Calyx deeply cleft, not enlarging after flowering; stamens
included; leaves oblong or lanceolate, obtuse. Hoary,
but corolla glabrous ... ... *Bowmanni*
Calyx deeply cleft, enlarged and membranous after
flowering.
 Leaves ovate, thick, hoary; stamens included; seeds
 velvety ... ... *rotundifolia*
 Leaves ovate-oblong, hoary; stamens exsert; seeds
 glabrous ... ... *leucophylla*
 Leaves linear or linear-lanceolate.

  (1.) *Stamens included.*

 Enlarged calyx-segments more or less cuneate
 and obtuse.
  Ovary shortly hairy; corolla small, pubes-
  cent outside ... ... *Paisleyi*
  Ovary woolly; leaves entire.
   Corolla small; leaves linear; sepals
   obovate ... ... *Sturtii*
   Corolla small; leaves channelled, ob-
   tuse; viscid ... ... *exilifolia*
   Corolla ¾ in., leaves linear-lanceolate;
   glabrous-viscid ... ... *Mitchelli*
 Ovary woolly; leaves serrate on the margin.
  Inner sepals lanceolate-oblong, outer
  ones ovate; viscid ... ... *Gibsoni*
 Enlarged calyx-segments lanceolate.
  Leaves linear, serrate.
   Pubescent; sepals linear-lanceolate ... *Berryi*
   Glabrous; sepals broadly lanceolate ... *Clarkei*

  Leaves linear, channelled; flower-stalks very
   long ... ... *Gilesii*
  Leaves cylindrical; flower-stalks very long *Hughesii*

 (2.) *Stamens, 2 or all, exceeding the corolla.*
  Corolla white or pinkish; leaves opposite or
   partly alternate; calyx-segments cuneate;
   ovary shortly hairy ... ... *oppositifolia*
  Corolla red; leaves scattered; calyx-segments
   lanceolate ... ... *Latrobei*

 II. Corolla very irregular, 4-upper lobes short,
   acute, the 5th deeply separate and narrow.
   Calyx-segments imbricate, enlarging after
   flowering.

Stamens exsert, fruit drupaceous.
 Flower-stalks shorter than the calyx; leaves lanceo-
  late, entire, somewhat hairy, rarely pubescent;
  flowers red, rarely green ... ... *Brownii*
 Flower-stalks longer than the calyx, flexuous-spread-
  ing.
  Leaves narrow-lanceolate, entire.
   Lowest corolla-lobe obtuse; corolla orange-
    red ... ... *Duttonii*
   Lowest corolla-lobe acute; corolla red, dark-
    spotted ... ... *maculata*
   Leaves lanceolate, usually serrate; calyx-
    segments lanceolate ... ... *denticulata*
   Leaves ovate; calyx-segments ovate, much en-
    larging; corolla green ... ... *latifolia*
Stamens included; fruit dry. Leaves narrow-linear;
 flowers large on long stalks; corolla rose-coloured,
 red-spotted ... ... *alternifolia*

Sub-Class V.—Gymnosperms.

ORDER CONIFERAE.

Callitris.

Cones-scales about 6 to 8 in 2 whorls. Cones globular;
 fruits numerous beneath each scale; leaves very
 minute; cotyledons 2 or 3. Trees.
Cone-scales 8, closely contiguous before expansion, each
 with a blunt subcentral protuberance, radially furrowed ... ... *verrucosa*
Cone-scales 6, separated by a slight furrow before expansion, each with a pointed prominence above the middle *cupressiformis*

ORDER CYCADEAE.
Encephalartos.

Leaves simple-pinnate, leaflets numerous, flat, 10- to 12-nerved; cone-scales large, cordate-reniform, pointed, glabrous ... ... *MacDonnelli*

CLASS II.—MONOCOTYLEDONS.

Sub-Class I.—Florideae Perigynae.

ORDER HYDROCHARIDEAE.
Ottelia.

Leaves radical on long stalks, the *lamina oval* or oblong, floating; flowers bisexual, large, solitary on long scapes, within a tubular 2-lobed wingless spathe; petals white, 3; stamens 6; stigmas 6, 2-lobed ... ... *ovalifolia*

Vallisneria.

Leaves radical, elongated, partially or wholly submerged, without lamina; flowers unisexual; female flower solitary, within a narrow tubular 3-toothed spathe, on a long spiral scape; petals 0; stigmas 3 ... ... *spiralis*

Blyxa.

Leaves long, grass-like, entire; male flowers within a tubular 2-toothed spathe, stamens 8; female flowers solitary within a long and slender spathe; petals present... ... *Roxburghii*

Hydrilla.

Leaves oblong-lanceolate, whorled along submerged much-branched stems; female flowers sessile, axillary, solitary, within a short tubular spathe; stigmas 3; petals present *verticillata*

Halophila.

Marine, leaves oval, long-stalked, submerged; female flowers singly sessile; stamens and stigmas 3; petals 0; flowers between distinct bracts ... ... *ovalis*

ORDER ORCHIDEAE.

Leaves reduced to scales. Root of rhizome-like tubers. Pollen masses waxy, attached to two stipule-like processes. Parasitic on roots ... ... **Dipodium**

Epiphytal. Stems enlarged into pseudo-bulbs. Pollen masses waxy, sessile on a gland ... ... **Cymbidium**

GENERA AND SPECIES.

Leaves developed. Terrestrial. Pollen granular.

  a. United stamens and style (column) short.

Flowers racemose, *sepals broad* and petal-like.
  Flowers regular, the labellum quite similar to the two other petals ... ... **Thelymitra**
  Labellum densely *hairy;* dorsal sepal concave, petals smaller ... ... **Calochilus**
Flowers racemose; *lateral sepals narrow-linear* and long; labellum 3-lobed, at or near its base.
  Dorsal sepal erect or spreading; lateral petals long ... ... **Diuris**
  Dorsal sepal concave, incurved over the column; lateral petals minute ... ... **Orthoceras**
Flowers racemose, *turned upside down;* sepals narrow, greenish.
  Leaves flat broad; flowers large ... ... **Cryptostylis**
  Leaf cylindrical; flowers often small; labellum callously thickened, usually with a broad thin margin ... ... **Prasophyllum**
Flowers spicate, small.
  Flowers *spirally arranged;* leaves narrow ... ... **Spiranthes**
  Flowers green; leaf cylindrical; upper sepal concave ... ... **Microtis**
Flower singly terminal; *labellum tubular* at the base ... **Corysanthes**

  b. Column elongate; leaves rarely more than one.

Leaves several. Dorsal sepal hooded connate with the lateral petals; labellum stalked; lower sepals much united forming a "lower lip" ... ... **Pterostylis**
Leaf one only; labellum sessile; dorsal sepal disconnected.
  Labellum smooth, with two adnate callosities at the base.
    Lateral petals much shorter than the sepals.
    Leaf cordate ... ... **Acianthus**
    Lateral petals about as long as the sepals.
    Leaf cordate ... ... **Cyrtostylis**
  Labellum smooth, with 1 or 2 clavate processes erect against the column; flowers almost regular.
  Leaf oblong ... ... **Glossodia**
  Labellum papillary, thick. Leaf cordate ... ... **Lyperanthus**
  Labellum densely hairy, very convex. Leaf narrow **Eriochilus**
  Labellum with glandular hairs in rows. Leaf narrow **Caladenia**

L

Dipodium.

Tall; calyx and petals almost equal, red-spotted; labellum bilobed ... ... *punctatum*

Cymbidium.

Leaves 6 to 12 inches long, keeled, channelled above ... *canaliculatum*

Thelymitra.

I. Column incurved *(hood)* over the anther, an appendage on each side terminated by a tuft of hairs.

  a. Petals blue or bluish.

Hair-tufts reaching beyond the hood; hood 3-lobed, the central one crested ... ... *ixioides*

Hair-tufts turned upwards not extending beyond the hood.
  Hood bilobed; the sinus narrow and deep, slightly denticulated; flowers about 7, pale-violet; stigma oblong-obcordate, anther contiguous. Rather tall, moderately stout ... ... *longifolia*
  Hood bilobed, the sinus short and broad, both without denticles; flowers 2 to 4, bluish-violet; stigma transversely round-oblong, anther distant. Slender, not tall ... ... *parviflora*

Hair-tufts horizontal.
  Hood bilobed, denticulate along the edges and the base of the sinus; flowers about 10, purplish-blue; anther much hidden behind the stigma. Rather tall ... ... *aristata*
  Hood bilobed and deeply denticulated, small toothed-lobe in the sinus; flowers about 30, large, greyish-blue; anther hidden by the stigma. Tall, robust *grandiflora*

  b. Petals yellow blotched with light-brown.

Hood deeply fringed with linear lobes, with a club-shaped dorsal appendage ... ... *fuscolutea*

  c. Petals bright-pink.

Hood hardly developed, 3-lobed; hair-tufts turned upwards, *yellow*, 2-lobed at the base. Rather tall, stout ... ... *luteocilium*

II. Column not hooded; lateral appendages without hair-tufts; stems flexuose, never tall.

  a. Petals yellow; sepals reddish outside.

Column produced in a terminal plate behind the anther.
  Terminal plate produced above the anther, undulate or almost denticulate; lateral appendages broad and rugose; column of an urn-like form; flowers 1 or 2 ... ... *urnalis*

Terminal plate shorter than the anther; lateral appendages orange-yellow, rough; flowers small, 1 or 2 ... ... *flexuosa*
Column without a plate; lateral appendages purplish, smooth, ovate- or obcordate-cuneate; flowers rather large, 2 or 1 ... ... *antennifera*

b. Petals and sepals red.

Terminal plate produced beyond the anther, slightly crenulate.
    Lateral appendages nearly smooth, bright-yellow; flowers dark-red, never expanding ... ... *carnea*
    Lateral appendages lanceolate, densely beset on both sides with rugose glands; flowers larger, bright-red, constantly expanding ... ... *rubra*

Calochilus.

Flowers several large, labellum fringed all over, sepals greenish ... ... *Robertsoni*

Diuris.

I. Labellum 3-lobed; the middle lobe with 2 raised longitudinal lines.

Petals lilac, middle lobe of labellum semiorbicular-rhomboid ... ... *punctata*
Petals yellow, dark-spotted. Lateral lobes of labellum equal or longer than middle lobe.
    Lateral sepals longer than the petals; leaves narrow-linear. Dwarf ... ... *palustris*
    Lateral sepals scarcely so long as petals; leaves broad-linear. Rather tall ... ... *maculata*
Petals yellow; lateral lobes of labellum less than half the length of the acute middle lobe, its raised lines pubescent ... ... *pedunculata*

II. Labellum deeply 3-partite, middle lobe with one raised line.

Petals yellow, purple-spotted; middle lobe of labellum rather acute, much exceeding the lateral lobes ... *sulphurea*
Petals yellowish mixed with brown, middle lobe of labellum dilated upwards hardly longer than the lateral lobes ... ... *longifolia*

Orthoceras.

Rather tall, rigid; flowers distant, large; bracts large ... *strictum*

Cryptostylis.

Rather tall, flowers 3 to 12, on very short stalklets; bracts large ... ... *longifolia*

Prasophyllum.

I. Labellum sessile. Usually tall and robust.

Lateral sepals connate to about the middle; fruit narrow.
  Labellum straight, its marginal portion narrow hardly wider than the callous portion; flowers comparatively large, yellowish-green. Tall and very robust ... *elatum*
  Labellum recurved from the middle, its thin *white* crisp portion much exceeding the callous portion; petals greenish, dark-streaked, pointed ... ... *australe*

Lateral sepals disconnected; fruit somewhat obliquely swollen.
  Labellum as in *P. elatum;* flowers dark-coloured or greenish ... ... *fuscum*
  Labellum as in *P. australe;* petals rather narrow, nearly blunt ... ... *patens*

II. Labellum stalked. Dwarf, slender.

Labellum acute, reddish, slightly denticulate; flowers very small, dark-reddish; petals narrow-lanceolate... *despectans*

Labellum rather blunt, extremely short; lateral sepals bulging at the base; flowers very small, dark-purplish and somewhat greenish; petals deltoid-lanceolate ... *nigricans*

Spiranthes.

Flowers red, the labellum white, numerous, often hairy   *australis*

Microtis.

Rather tall and stout; flowers light-green, very small; lateral sepals recurved; labellum with a tubercle near the end ... ... *porrifolia*

Dwarf, very slender; flowers yellowish-green, minute, drying black; lateral sepals spreading *(M. atrata)*... *minutiflora*

Corysanthes.

Dwarf; leaf one, orbicular-cordate; flower large, dark-purple ... ... *pruinosa*

Pterostylis.

I. Lower sepals erect. Hood green.

Leaves in a radical rosette.
  Labellum shortly and broadly bilobed; leaves small, ovate. Dwarf ... ... *concinna*
  Labellum entire. Flowers about ½ inch; lobes of the lower lip separate by a wide sinus with an inflexed tooth.
    Labellum linear elliptical; leaves orbicular. Dwarf ... ... *nana*

Labellum entire. Flowers 1 inch or more.
  Lobes of the lower lip lanceolate, with an acute sinus between them.
    Flowers on long stalklets, much bent downwards, so as to appear nodding ... ... *nutans*
    Flowers erect on long stalklets ... ... *pedunculata*
  Lobes of the lower lip separated by a wide sinus *curta*
Stems leafy.
  Leaves crowded at the base of the stem passing gradually into stem leaves or scales. Labellum hardly pointed. Rather tall; flower large ... *cucullata*
  Stem-leaves few, linear; labellum pointed; hood short and slightly incurved. Rather dwarf, slender ... ... *praecox*
  Lower leaves reduced to scales passing up into lanceolate leaves; labellum pointed; hood elongate and much incurved ... ... *reflexa*
  Stem-leaves narrow-lanceolate, labellum quite blunt *obtusa*

  II. Lower sepals reflexed or recurved from the middle.

Flower large, solitary; labellum linear-cylindrical, beset with yellow hairs, ending in a small glabrous dilatation ... ... *barbata*
Flowers two or more in the raceme.
  Leaves in a radical rosette.
    Calyx-lobes obtuse; lower lip shortly 2-lobed; flowers small ... ... *mutica*
    Calyx-lobes with fine points; lip deeply 2-lobed; flowers large ... ... *rufa*
  Stems leafy, no rosette; flowers large.
    Hood banded with narrow red lines; labellum rough; leaves linear; column abruptly dilated upwards, somewhat fringed ... ... *longifolia*
    Hood banded with broad red lines; labellum slightly fringed, with a semi-lanceolate minute appendage; leaves lanceolate; column gradually expanded towards the middle ... ... *vittata*

Acianthus.

Dorsal sepal very much elongated; flowers dark purple. Dwarf ... ... *caudatus*
Dorsal sepal quite short; flowers pale pink. Dwarf, slender ... ... *exsertus*

Cyrtostylis.

Flowers small, purple; callosities of the labellum dark-red ... *reniformis*

Glossodia.

Labellum-appendage short, bilobed; flowers one or two, large, bluish or lilac, rarely white inside, paler coloured outside ... *major*

Lyperanthus.

Flowers 2 to 4, purple, large. Somewhat dwarf, but robust, drying black ... *nigricans*

Eriochilus.

Labellum ovate-cuneate, much recurved, slightly fringed; flowers 1 to 3, small, pinkish; leaf cordate- to lanceolate-ovate ... *autumnalis*

Labellum semiorbicular-cuneate, nearly flat, conspicuously fringed; flowers 1, sometimes 2, reddish; leaf lanceolate-ovate; one elliptical leafy bract usually on the stem ... *fimbriatus*

Caladenia.

1. Labellum with divergent forked veins or colour-lines; petals about as long as the sepals; flowers 1, or rarely 2.

Labellum with an entire margin, orbicular-ovate, shortly stalked; calli in 2 rows; petals linear; sepals narrow-lanceolate, shortly acuminate, about $\frac{1}{2}$ in. ... *Cairnsiana*

Labellum fringed or toothed on the margin.
    Labellum ciliate-fringed, broadly ovate with a lanceolate apical extension, shortly stalked; calli in 4 rows; sepals lanceolate, tapering to a clavate point, 1 in. or more ... *reticulata*

    Labellum toothed or serrate.
        Labellum crescent-shaped with a broad apical extension, long-stalked, anterior margin with pointed denticulations; calli clustered or obscurely 4-rowed; sepals lanceolate, pointed, $\frac{1}{2}$ in. or more; leaf oblong-lanceolate ... *toxochila*

        Labellum narrowly rhomboid-ovate, sessile, anterior margin bluntly serrate; calli in 2 rows; sepals subulate, *densely glandular-hairy*, nearly 2 in. long; leaf linear ... *tentaculata*

II. Labellum without forked veins.
    a. Petals much longer than the sepals.

Petals erect, narrow-linear, clavate towards the end; flowers 1 or 2, on long stalks; calli in 2 or 4 rows ... *Menziesii*

    b. Petals not exceeding the sepals.

Sepals with long tapering points, dorsal sepal erect and much incurved; flowers 1 or 2.
  Calli in 2 rows; flowers red, sepal-points very long and thread-like ... ... *filamentosa*
  Calli in 4 rows.
    Labellum very broad; the lateral lobes yellow, deeply dissected on the margin; middle lobe ovate, purple; sepals and petals with long rapidly tapering points ... ... *dilatata*
    Labellum ovate, purplish or whitish; the lateral lobes hardly prominent, shortly fringed; sepals and petals gradually tapering from a broad base ... ... *Patersoni*
    Labellum oblong, dark-red, minutely denticulated towards the base ... ... *leptochila*

Sepals with short points, dorsal sepal erect and concave.
  Leaf oblong-lanceolate; flowers usually 3, white or pink; labellum deeply 3-lobed; calli in 2 short rows ... ... *latifolia*
  Leaf narrow-linear, labellum slightly lobed or almost entire.
    Flowers pink or white, 1 to 5; calli in 2, or rarely 4, rows, yellow or red; labellum slightly trifid and fringed ... ... *carnea*
    Flowers blue, rarely white, solitary.
      Labellum slightly trifid; calli in 2 rows, yellow ... ... *coerulea*
      Labellum almost entire, denticulate-fringed; calli crowded, blue ... ... *deformis*

ORDER IRIDEAE.

Calyx-lobes petaloid, blue or rarely white, much larger than the petals. Style longer than stamens (3), with 3 broad stigmas. Flowers in solitary terminal spikes on leafless scapes; filaments united below ... **Patersonia**

Calyx-lobes petaloid, blue, nearly equal to the petals. Style shorter than the stamens (3), with 3 linear stigmas; filaments almost free ... ... **Sisyrinchium**

Patersonia.

Scape short; outer bracts of spikes striate; calyx-tube filiform, somewhat exsert ... ... *glauca*

Scape longer than the leaves; outer bracts almost smooth; calyx-tube slender, enclosed ... ... *longiscapa*

Sisyrinchium.

Spikes with several flowers, the outer bracts broad with scarious margins ... ... *cyaneum*

ORDER AMARYLLIDEAE.

Flowers solitary or in a few-flowered raceme, stamens free; calyx-lobes green outside, yellow within; petals yellow ... ... **Hypoxis**

Flowers umbellate; calyx and petals white or yellow; stamens free ... ... **Crinum**

Flowers umbellate, calyx petaloid; filaments united into a wide tube; flowering scapes appearing before the leaves ... ... **Calostemma**

Hypoxis.

Leaves beset with long soft hairs; flowers 1 to 5 on the scape; anthers much divergent at the base; capsule obovoid-globular ... ... *hygrometrica*

Leaves glabrous; anthers parallel; flowers usually solitary.
    Flowers large, scape with a large bract about the middle; fruit ovoid... ... ... *glabella*
    Flowers small; scape with 2 opposite bracts; fruit ovoid-globular ... ... *pusilla*

Crinum.

Flowers many, sessile in the umbel, or on stalks shorter than the beaked ovary; leaves long, but narrow... *angustifolium*

Flowers on pedicels longer than the obtuse ovary.
    Flowers primrose-yellow, 6 to 12 in the umbel, the lobes about 3 in. long and nearly 1 in. broad ... *flaccidum*
    Flowers yellowish-white, 4 in the umbels, lobes smaller ... ... *pedunculatum*

Calostemma.

Calyx-tube dilated; flowers purple ... ... *purpureum*

Flowers yellow, larger about ½ inch long ... ... *luteum*

GENERA AND SPECIES.

Sub-Class II.—Florideae Hypogynae.
ORDER LILIACEAE.

I. Style 3-cleft.

Flowers in a terminal umbel; sepals free, petaloid; root fibrous ...	Burchardia
Flowers spicate, mostly unisexual; sepals petaloid, somewhat connate with the petals; root bulbous ...	Wurmbea
Flowers clustered in interrupted spikes or at end of scapes or paniculate branches or in racemes; unisexual, often in separate plants. *Leaves firm*, densely tufted; sepals and petals often connate at the base ...	Xerotes

II. Style undivided.

Flowers in panicles, blue; *anthers opening by pores*; fruit succulent; stems leafy; roots fibrous, the stock often branched ...	Dianella
Flowers racemose or paniculate; sepals and petals alike; fruit dry.	
Petals and sepals deciduous, yellow; filaments bearded ...	Bulbine
Petals and sepals persistent, spirally twisted over the fruit after flowering, blue; filaments glabrous	Caesia
Petals and sepals persistent, purplish, not twisted after flowering, filaments hairy or the anthers with basal crests ...	Arthropodium
Flowers umbellate or paniculate; sepals and petals alike, *persistent, twisting after flowering*.	
Petals fringed; capsule lobeless, bursting; flowers purple ...	Thysanotus
Petals fringeless; fruit of 3, 1-seeded indehiscent fruitlets ...	Tricoryne
Flowers in loose dichotomous cymes; sepals and petals alike, twisted after flowering; capsule 3-lobed; flowers blue ...	Chamaescilla
Flowers small in clusters; sepals and petals white or pale-reddish; capsule dry; root fibrous ...	Laxmannia
Flowers solitary, terminal; branches leafy; sepals and petals alike, connate towards the base, blue; anthers opening by pores; *fruit indehiscent*, 1-seeded	Calectasia
Flowers solitary or rarely 2 together along the branchlets; sepals and petals whitish, twisted after flowering; fruit indehiscent 1-seeded; anthers opening by terminal pores ...	Ccrynotheca

Flowers very numerous in dense cylindrical terminal spike; sepals glume-like; petals membranous with white spreading tips. More or less arborescent and palm-like; leaves very long, rigid, sharp-pointed ... **Xanthorrhoea**

Burchardia.

Leaves few, narrow-linear, concave; flowers white, tinged with pink ... ... *umbellata*

Wurmbea.

Leaves few, filiform; flowers white or pink, with a dark band, few... ... ... *dioica*

Xerotes.

I. Male flowers sessile in clusters of a whorled panicle; capsule smooth.

Bracts narrow, often elongate and pungent; sepals free, brown; petals shortly united, yellow; leaves mostly *2-toothed* at the apex ... ... *longifolia*

Bracts obtuse and short; sepals free; corolla cleft; *3 stamens adnate to the centre of the corolla-lobes*, 3 alternate adnate to tube ... ... *dura*

II. Male flowers stalked, clustered in simple whorls.

Bracts small, very short; petals yellow; capsule wrinkled ... ... *Brownii*

III. Male flowers scattered in racemes or panicles.

Flowers *almost sessile*, comparatively large, yellow; leaves 2-toothed at the apex; capsule longitudinally striate. Panicle spreading ... ... *effusa*

Flowers minute, conspicuously stalked; leaves very narrow or almost filiform; panicle narrow or reduced to a single raceme. Capsule smooth.

Petals and sepals greenish-yellow or brownish, equal, very spreading ... ... *micrantha*

Petals yellow, rather thick, ovate; sepals greenish, thinner and shorter ... ... *Thunbergii*

IV. Male flowers in globular clusters, terminal or in interrupted spikes.

Leaves on the stems, as well as radical.

Petals bright-yellow, male flowers in spikes; capsule slightly wrinkled; leaves under 6 in., sometimes slightly twisted; female flowers in sessile heads ... ... *glauca*

Male flowers on a short, simple or branched scape; leaves 1 foot long; female heads shortly stalked; bracts scarious long-pointed ... ... *elongata*

Leaves radical, or nearly so, 1 to 2 feet; scapes shorter, with flower-heads, terminal, scattered or spicate; petals white; capsule smooth ... ... *leucocephala*

Leafless except sheathing scales; root creeping; scapes tufted, a few fertile with terminal heads; barren scapes rush-like, pointed; capsule 3-furrowed, smooth ... ... *juncea*

Dianella.

Stems almost leafless; leaves long, comparatively narrow, smooth at the margin.
 Anthers almost black; leaves rigid with revolute margins; clasping leaf-stalks closed, keeled; berry black, globular ... ... *revoluta*
 Anthers yellow; leaves flat, sheathing base quite open at the summit; berry white, globular ... *laevis*

Bulbine.

Scapes leafless; flowers racemose; filaments beset with hairs ... ... *bulbosa*
 Three of the filaments quite glabrous; flowers smaller ... ... *semibarbata*

Caesia.

Flowers blue, somewhat pendulous, irregularly paniculate; leaves broadly linear, lax, mostly radical ... *vittata*
Flowers paler, smaller, in racemes; leaves narrowly linear ... ... *parviflora*

Arthropodium

Filaments hairy, glabrous towards the base only.
 Stems 1 to 2 feet; leaves broad-linear; flowers 2 to 4, on each pedicel; anthers elliptic; seeds several; filaments hairy only above the middle *paniculatum*
 Stems shorter, leaves narrower, pedicels 1-flowered; flowers smaller, anthers ovate, seeds few; filaments hairy nearly to the base ... ... *minus*
Filaments glabrous, the anthers with 2 small crest-like appendages at base.
 Capsule on erect stalks; anther-appendages very short; sepals obtuse and somewhat crisped ... *strictum*
 Capsule on reflexed stalks; anther-appendages rather long; sepals obtuse and somewhat fringed *fimbriatum*

Thysanotus.

Stem erect, much branched from near the base; rigid, terete, striate; basal leaves linear or wanting; lower branches often fruitless; flowers few in an umbel at the end of the branches ... ... *dichotomus*

Stem erect, unbranched in the lower portion.
  Flowers in a loose panicle; stamens 3 short and 3 long ... ... *tuberosus*
    Leaves long; stamens all equal in length ... *exasperatus*
  Flowers few, in umbels, or several sessile along the upper part of the scape, with *broad* white *bracts* *Baueri*

Stem twining branched; basal leaves few, upper leaves minute or wanting; flowers solitary or 2 together, terminal ... ... *Patersoni*

Dwarf, branches slender, intricate, flowers exceedingly small ... ... *exiliflorus*

Tricoryne.

Stems wiry, terete, with clustered branches; flowers yellow, in umbels ... ... *elatior*

Chamaescilla.

Leaves often shorter than the scape; flowers few on long pedicels ... ... *corymbosa*

Laxmannia.

Stems branched, filiform; leaves tufted at the base of the branches and under the sessile flower-heads; sepals and petals transparent ... ... *sessiliflora*

Calectasia.

Stems clustered, firm; leaves crowded, linear, very acute or sharp-pointed ... ... *cyanea*

Corynotheca.

Slender, rigid, dichotomously branched; leafless ... *lateriflora*

Xanthorrhoea.

1. Stems arborescent.

Leaves quadrangular; trunks about 5 or 6 ft.; usually simple. Scapes 3 to 6 feet ... ... *quadrangulata*

Leaves flat below, slightly convex above; trunks up 15 feet, often divided; scapes up to 20 feet ... *Tateana*

II. Stems very short, trunk-less.

Leaves flat or more or less triquetrous, spikes under
1 foot long; floral bracts spathulate ... ... *minor*

Leaves flat, slightly convex above; spikes up to 6 feet
long; floral bracts narrow, acuminate ... ... *semiplana*

ORDER XYRIDEAE.
Xyris.

Tufted perennials; leaves radical, more or less grass-like;
flowers in a terminal head, within imbricate rigid bracts;
scape long.

Herbaceous sepals opaque with scarious margins, prominently
keeled, usually ciliate at the end; *bracts* irregularly
arranged *in 5 rows*; placentas united at the base, as long
as the ovary ... ... *operculata*

Herbaceous sepals shining, entire, hardly keeled; placentas
much shorter than the ovary, almost distinct ... ... *gracilis*

ORDER COMMELINEAE.
Commelina.

Flowers blue within a spathe. Perfect stamens 3, staminodia
3; sepals and petals distinct. Procumbent; leaves
linear-lanceolate; spathe oblique, funnel-shaped... ... *ensifolia*

ORDER ALISMACEAE.
Damasonium.

Fruitlets 2-seeded, beaked. Semiaquatic herb. Leaves basal,
on long stalks, from ovate-cordate to lanceolate. Flowers
singly terminal or umbellate at the end of whorled
pedicels ... ... *australe*

ORDER JUNCACEAE.

Fruit 3-seeded; leaves grass-like, chiefly radical, hairy... **Luzula**
Fruit many-seeded; leaves grass- or rush-like, glabrous **Juncus**

Luzula.

Tufted perennial; flowers in umbellate or crowded clusters *campestris*

Juncus.

I. Leaves grass-like. Inflorescence terminal with
spreading bracts.

Stems branchless; leaves radical, flaccid; flowers brown.
Leaves broad-linear; flower-clusters in unequally
compound spreading cymes. Stamens 3 ... ... *planifolius*

Leaves narrow-linear; flower-clusters cymous, stamens 6 ... *caespititius*

Stems branched; flowers pale-coloured, usually not clustered, in a branched leafy panicle. Stamens 6, rarely 3. Dwarf annual... *bufonius*

  II. Leaves channelled, almost cylindrical.

Flowers clustered in a slightly branched leafy cyme; stamens 6. Dwarf tufted perennial ... *homalocaulis*

  II. Leaves or stems cylindrical. Panicles appearing lateral, by the subtending leafy-bract continuing the stem.

Filaments filiform, seeds not appendaged.
 Leaves almost all reduced to sheathing scales.
  Tall, stout; leaf-scales long; flowers pale-coloured, stamens usually 3; leafy bract erect, pungent ... *pallidus*
  Rather tall, slender; flowers usually few, dark-coloured; stamens 6 ... *pauciflorus*
  Tall, stout; flowers numerous, dark-coloured; stamens 3; leaf-scales short; panicles scattered or densely clustered or head-like ... *communis*
 Leaves several, scattered; stems compressed, jointed; flower-clusters numerous; stamens usually 6 ... *prismatocarpus*

Filaments flattened; seeds appendaged at both ends; flower-clusters in an irregular compound cyme; leafy bract long, erect, pointed. Stamens 6 ... *maritimus*

ORDER PALMAE.

Livistona.

Erect, with a terminal crown of fan-shaped leaves; flowers bisexual; sepals free, petals valvate, connate at the base; stamens 6, filaments very broad at the base, but free ... *Mariae*

ORDER TYPHACEAE.

Typha.

Stems about 4 ft.; leaves often as long; upper spike separated by a short interval from the lower; stamens 2 or 3, connate. Ovary of a single carpel; stigma unilateral ... *angustifolia*

ORDER FLUVIALES.

 I. Flowers clustered, in spikes or racemes.

Sepals and petals present; flowers bisexual, in spikes.
 Fruitlets 3 or 6, coherent till ripe; leaves radical; sepals and petals usually 3, bract-like. Aquatic or terrestrial ... **Triglochin**

Fruitlets 4; branches leafy; sepals 2; petals 2, small, bract-like. Aquatic plants, leaves usually stipulate ... ... **Potamogeton**

Sepals present, 3, bract-like; flowers bisexual.
Fruitlet one; branches leafy, flowers in spikes. Maritime ... ... **Posidonia**

Sepals absent; flowers in spikes.
Fruitlets usually 4; flowers bisexual, spikes on long spiral filiform stalks; stamens 2. Aquatic. **Ruppia**

Fruitlet solitary. Flowers of both sexes in 2 alternate rows, enclosed in the sheathing base of floral leaves; stamen of one anther; stigmas 2. Maritime ... ... **Zostera**

II. Flowers solitary, scattered, unisexual within sheathing bracts.

Fruitlets 3, stalked; stamens 3, anthers connate; stigma 1; leaves capillary, alternate. Calyx 3-lobed. Aquatic ... ... **Lepilaena**

Fruitlets 2; style bifid; anthers 2, sessile, connate, 4-celled; flowers concealed by the clasping appressed leaf-sheaths; leaves alternate. Maritime... ... **Cymodocea**

Fruit simple, often connate with the small tubular calyx; leaves opposite, narrow; stigmas 2 to 4. Aquatic ... ... **Naias**

Triglochin.

I. Fertile fruitlets 3, separating when ripe from a central axis.

Dwarf tufted terrestrial annuals; leaves filiform.
Fruitlets linear, 3-ribbed, bidenticulate at the base; lower flowers often with 3 sepals and 1 stamen only, terminal flower with 6 stamens ... ... *centrocarpa*

Fruitlets broadly ovoid and terminated by the spreading style; lower flowers mostly with 1 stamen only, terminal flower with 3 stamens ... *mucronata*

Semiaquatic, somewhat tall and slender, stoloniferous; leaves filiform; flowers generally numerous, with 3 stamens; fruits orbicular, compressed, dorsally streaked ... ... *striata*

II. Fertile fruitlets usually 6, orbicular to narrow-oblong, often twisted; no central axis.

Aquatic; leaves broadly linear, very long, upper part floating ... ... *procera*

Potamogeton.

Leaves dissimilar; floating leaves firm, on long stalks; submerged leaves membranous alternate.
    Floating leaves 2 to 4 in., oval, subcordate, about 20-nerved; fruit ovoid, 3-angled on the back, shortly beaked ... ... *Tepperi*
    Floating leaves elliptical or lanceolate, about 1 in., few-nerved; fruit distinctly beaked ... ... *tenuicaulis*

Leaves all submerged, flat, membranous, simply sessile.
    Leaves undulate-crisped on the margin, narrow-oblong, blunt, with a strong central nerve and one on each side of it ... ... *crispus*
    Leaves narrow-linear, obtuse, 1-nerved; stipules blunt; flower-spikes about ½ in. long, of several flowers ... ... *ochreatus*
    Leaves narrow-linear, acute, 1- or 3-nerved; stipules acute; spikes short, few-flowered ... ... *acutifolius*

Leaves all submerged, mostly dilated and clasping at the base. Stems filiform, dichotomously branching; leaves narrow-linear, alternate... ... ... *pectinatus*

Posidonia.

Stems branched; leaves very long, broad-linear, rounded at the end ... ... *australis*

Ruppia.

Stems and branches very slender; leaves capillary, long, clasping ... ... *maritima*

Zostera.

Each flower subtended by a transverse vertical bract, flowers few; leaves narrow-linear, truncate or notched at the end, up to 1 or 2 feet long, the sheathing base rather short ... ... *nana*

Flowers several, without bracts; leaves rounded at the at the end, sheathing base about one inch long ... *Tasmanica*

Lepilaena.

Stems filiform, much-branched; styles longer than the carpels; flower-stalks very short; calyx-lobes rather longer than the carpels ... ... *Preissii*

Styles much shorter than the carpels; flower-stalks lengthening to about ½ inch; calyx-lobes ovate, very short ... ... *australis*

Cymodocea.

Stems and branches hard; leaves firm, broad-linear, truncate or toothed at the end, 1 to 3 inches long   *antarctica*

Naias.

Stems slender; leaves very narrow-linear, about 1 in., minutely toothed, the sheathing base produced into broad stipules ... *tenuifolia*

Leaves linear, prominently toothed; no stipules ... *major*

ORDER LEMNACEAE.

Fronds emitting capillary roots; flower in a fissure of the margin, supported by a bract; anther 2-celled ... **Lemna**

Fronds minute, without roots; flower in a cavity on the upper side, no bract, anther 1-celled ... **Wolffia**

Lemna.

Root-fibres one to each frond.
  Fronds very thin, oblong-lanceolate, the young ones often projecting cross-like at both ends... *trisulca*
  Fronds broadly ovate, convex underneath; about 2 lines long ... *minor*
Root-fibres 2 to 5 to each frond. Fronds thin, oval or oblong, rather larger than *L. minor* ... *oligorrhiza*

Wolffia.

Fronds ovate, about ½ line diameter, but very convex underneath ... *Michelii*

ORDER RESTIACEAE.

I. Minute tufted bisexual plants; leaves radical linear; flowers comprised ordinarily of one sepal, one stamen, and one ovary.

Flowers in a depressed head-like cluster, surrounded by spreading transparent bracts; floral bracts 0; ovary 1-celled, stigmas 2 or 3 ... **Trithuria**

Flowers in a single spike on a slender scape; fruit 1-celled, opening by a slit.
  Spike supported by several bracts in 2 rows; each flower with 1 or 2 hyaline scarious sepals ... **Aphelia**
  Spikes supported by 2 sheathing bracts, each flower with 1 to 3 hyaline scarious sepals ... **Centrolepis**

II. Rush- or sedge-like, mostly unisexual, plants; sepals and petals 2 or 3 each; carpels 3, united into a single pistil; fruit 1- to 3-celled; stamens 3.

Fruit two- or three-celled; stigmas 2 or 3.
  Male and female inflorescence in spike-like panicles; bracts loosely imbricate, 2 bracteoles under each flower ... **Lepyrodia**

M

Flowers in spikelets, in both sexes nearly similar; bracts closely imbricate; no bracteoles ... ... **Restio**

Fruit 1-celled, 1-seeded; stigmas 3.

Inflorescence in both sexes in several-flowered spikelets, usually paniculate; fruit opening laterally ... **Leptocarpus**

Female spikelets 1-flowered; fruit indehiscent; male flowers several in spikelets or paniculate ... ... **Calostrophus**

Fruit 1-celled, 1-seeded; style undivided.

Male and female flowers several together in spikelets, often terminal; bracts imbricate ... ... **Lepidobolus**

Trithuria.

Bracts lanceolate, obtuse; heads singly terminal on slender scapes ... ... *submersa*

Aphelia.

Spike ovate, reclining; bracts with a narrow membranous margin... ... ... *gracilis*

Spike ovate, erect; bracts broadly membranous at the margin ... ... *pumilio*

Centrolepis.

I. Outer bracts glabrous.

Spikelet very narrow, of a reddish hue, containing several flowers, only the lowest male; pistils coherent; outer bract with a rigid recurved awn ... ... *polygyna*

Spikelet rather broad, containing several flowers; outer bract short with a short awn ... ... *glabra*

Spikelet ovate, containing many flowers; outer bracts with a long leafy point ... ... *aristata*

II. Outer bracts hairy, spikes ovate.

Outer bracts somewhat appressed, with long points; carpels 3 ... ... *fascicularis*

Outer bracts spreading, with short points; carpels 6 ... *strigosa*

Lepyrodia.

Stems tall, sheathing scales appressed; sepals as long or longer than the petals, both long and acute; bracts obtuse ... ... *Muelleri*

Restio.

Stems much compressed, simple erect; sheathing scales appressed, occasionally developing short laminae; spikelets in racemes ... ... *complanatus*

Stems cylindrical, very tall, branched; the sterile branches bearing numerous minute clustered leaves; spikelets in panicles ... ... *tetraphyllus*

Leptocarpus.

Male spikelets small, numerous, dark brown, in a terminal panicle; female spikelets few in a cylindrical spike; bracts ovate, obtuse; stems greyish ... ... *tenax*

Male spikelets large, few, almost ellipsoid, rich-brown on short filiform stalks; female inflorescence in a short spike-like panicle; bracts acutely acuminate; stems pale-green ... ... *Brownii*

Calostrophus.

Spikelets axillary all solitary and sessile; male spikelets 2- to 4-flowered; leaves minute; stems slender, lax much branched ... ... *lateriflorus*

Spikelets terminal; male spikelets in loose panicles; female spikelet comparatively large, solitary or rarely 2 or 3 together; leaves rudimentary; stems slender, much branched ... ... *fastigiatus*

Lepidobolus.

Stems simple, straight or flexuose; bracts oblong, acuminate with a short point; sepals ciliate; petals 3, narrower ... ... *drapetocoleus*

Sub-Class III.—Glumiferae.

ORDER CYPERACEAE.

1. Floral bracts in 2 straight rows (distichous).

Fruit only one in each spikelet. (Also some species of *Schoenus*).
 Hypogynous scales present; spikelet solitary, terminal, with 2 flowers, only one female; stamens 3; stigmas 3 ... ... **Lepidospora**

 No hypogynous scales or bristles; spikelets with 1 or 2 flowers, only one female, clustered in a dense globular head within an involucre of leafy bracts **Kyllingia**

Fruits more than one in each spikelet.
 Spikelets several-flowered, bisexual; one or two of the lowest bracts empty. Inflorescence spicate or umbellate with involucral bracts; no hypogynous scales or bristles; stigmas 2 or 3 ... ... **Cyperus**

 Spikelets 2- to 6-flowered, all fertile or the uppermost sterile, several outer bracts empty. Spikelets solitary or in capitate or paniculate clusters. Hypogynous bristles or scales present or wanting. Stigmas 3 ... ... **Schoenus**

II. Floral bracts in spiral rows (imbricate all round).

Fruits more than one in each spikelet.
- Base of style enlarged.
    - Style-base jointed on the fruit, deciduous. Spikelets clustered or umbellate. No hypogynous bristles ... ... **Fimbristylis**
    - Style-base continous with the fruit, persistent. Spikelets solitary, terminal. Hypogynous bristles. *No true leaves* ... ... **Heleocharis**
- Style filiform throughout.
    - Hypogynous bristles present. Spikelets solitary, or clustered, or umbellate, often lateral. Tall, stout plants ... ... **Scirpus**
    - No hypogynous bristles; usually small, slender **Isolepis**
    - Hypogynous scales 2, flat; spikelets in a terminal head ... ... **Lipocarpha**
    - No hypogynous bristles, fruit enclosed in a utricle.
        - Flowers strictly unisexual, either the sexes in separate spicate clusters or separately aggregated in the cluster ... ... **Carex**

Fruit only one in each spikelet; flowers 2 or few, only 1 fertile.
- *Branches leafless*, excepting sheathing scales as in Restiaceæ; spikelets solitary terminal. No hypogynous scales or bristles; stamens usually 5; stigmas 3 ... ... **Caustis**
- No hypogynous bristles or scales; stamens 3 to 6; stigmas 3 ... ... **Cladium**
- Hypogynous scales 6, in 2 rows, thickened, acuminate, adnate to the fruit. Stamens 3; stigmas 3. Leaves radical ... ... **Lepidosperma**
- *Stamens 6 to 12, each subtended by a scale;* bracteoles 2, opposite, navicular, ciliate on the keel. Inflorescence globular ... ... **Chorizandra**

Lepidospora.

Stems slender, leaves radical very narrow, spikelet without bracts ... ... *tenuissima*

Kyllingia.

Larger bracts nearly equal; fruit much smaller than the bract ... ... *monocephala*
Larger bracts very unequal; fruit as long as the bract *intermedia*

GENERA AND SPECIES.

Cyperus.

I. Spikelets flat with *navicular keeled bracts*; fruit *biangular*, stigmas 2.

Spikelets few in a loose cluster; stamens 2 ... ... *Eragrostis*

II. Spikelets flat; *rhachis not winged*; stigmas 3, fruit triangular.

Spikelets spreading, pale-coloured, in a *single* sessile cluster; bracts obtuse.
    Dwarf annual; spikelets 1 or 2, one long involucral bract; stamens 1 or 2... ... ... *tenellus*
    Slender perennial; spikelets 1 or few, involucral bracts 1 to 3; stamens 3; bracts greenish, 3- to 4-nerved ... ... *gracilis*
    Dwarf annual, spikelets numerous, involucral bracts 2 or 3 with a broad base, stamen 1 ... *pygmaeus*

Spikelets numerous, capitate or in an umbel of few rays. *Bracts* with a *prominent* straight or recurved *point* *squarrosus*

Spikelets dark-coloured in dense globose heads ... *difformis*

Spikelets pale or brown, capitate, or *solitary* on the rays of an umbel.
    Involucral bracts few unequal ... ... *trinervis*
    Involucral bracts 6, rigid, nearly equal ... ... *vaginatus*

Spikelets pale- or dark-brown, *clustered* on the *rays* of of an umbel.
    Spikelets small, few-flowered, in little globular clusters ... ... *holoschoenus*
    Spikelets linear; bracts tipped with fine points ... *Gilesii*
    Spikelets rather thick; bracts obtuse, or scarcely acute.
        Stems obtusely triangular; spikelets 8- to 12-flowered, linear-lanceolate, very spreading, of a golden-brown; bracts 2- or 3-nerved ... *fulvus*
        Stems acutely triangular; spikelets 10- to 30-flowered, linear, pale-brown; bracts 3- to 4-nerved. Involucral bracts rough ... ... *alterniflorus*

Spikelets pale-brown or yellowish-green, numerous in loose spikes along the rays of a simple or compound umbel; bracts very obtuse ... ... *Iria*

III. Spikelets flat or *round*, rhachis winged; stigmas 3; nut triangular.

Spikelets clustered in short spikes or umbels; nut not half the length of the bract.
    Stems leafless; bracts keeled, several-nerved ... *diphyllus*

Stems leafy at the base.
>> Spikelets scarcely flattened, very narrow, in dense clusters ... *subulatus*
>> Spikelets rather flat, 6 to 10, in loose clusters *rotundus*
> Spikelets flat in simple or compound umbellate spikes; nut usually as long as the bract. Tall ... *lucidus*
> Spikelets very flat in lengthened spikes, along the rays of a compound umbel; nut much shorter than the bract. Tall ... *exaltatus*

Schoenus.

I. Stems leafless, except sheathing scales at the base; no hypogynous bristles.

Stems thread-like, very weak, about 1 ft, ... *capillaris*
Stems rush-like, tufted, 6 to 10 in.; spikelets forming a single terminal head... ... *aphyllus*
Stems rush-like from a creeping rhizome, 1 to 2 ft.; spikelets in a narrow panicle... ... *brevifolius*

II. Leaves developed on the stems and at the base, flaccid; stamens 3.

Spikelets black, few in a terminal cluster, with a few axillary ones lower down; hypogynous bristles 6; flowers usually 2 in a spikelet; nut 3-ribbed. Rather dwarf ... *apogon*
Spikelets 1 or 2 together, mostly axillary; each producing one smooth, 3-ribbed fruit; hypogynous bristles 6 or fewer. Quite dwarf ... *axillaris*
Spikelets several together, axillary; hypogynous bristles 0, or rarely 2 or 3; nut deeply pitted or cancellate. Dwarf... ... *sculptus*
Stems submerged, branched; leaves filiform; spikelet solitary, terminal; no hypogynous bristles ... *fluitans*

III. Leaves at the base only.

Very tall, massively tufted; spikelets forming a large terminal head; flowers 2, but only 1 fertile in each spikelet; no hypogynous bristles ... *sphaerocephalus*
Dwarf not exceeding 1 ft.; spikelets in a single head, the erect involucral bract continuing the stem.
> Hypogynous bristles ciliate at the base; stems from a creeping rhizome; leaves terete, furrowed *nitens*
> Bristles densely hairy; leaf-sheaths densely *bearded* at the orifice and with short subulate *laminae*; stems tufted ... *deformis*

Minute plants, about 2 in., densely tufted; spikelets solitary, on stalks shorter than the leaves; no bristles.
    Leaves firm, channelled; fertile flower 1; nut obovate, rugose ... ... *Tepperi*
    Leaves flat, streaked; fertile flowers 2; nut ovoid, smooth, raised on a thin disk ... ... *discifer*

Fimbristylis.

Nut longitudinally and transversely striate; style ciliate, stamens usually 3 ... ... *communis*
Nut almost smooth, minutely striate; style ciliate.
    Small tufted annual, stems filiform; leaves linear, hairy; stamen 1 ... ... *velata*
    Tall, glabrous; stems rigid; leaves narrow-linear; stamens 3 ... ... *ferruginea*
Nut granular; leaf-sheaths ciliate at the orifice; style glabrous; stamens usually 1 ... ... *barbata*
Nut tuberculate, 3-angled; style glabrous; stamens 3   *Neilsoni*

Heleocharis.

Stems round, hollow, partitioned, up to 5 ft.; bracts dark-coloured with a hyaline border ... ... *sphacelata*
Stems solid, somewhat darf; bracts *keeled*.
    Stems rather slender, round.
        Sheathing scales with a small erect point; nut biconvex; stigmas 2 ... ... *acuta*
        No point to the scales; nut triangular; stigmas 3 ... ... *multicaulis*
    Stems filiform; hypogynous bristles few; nut triangular striate ... ... *acicularis*

Scirpus.

Spikelets 3 to 6, sessile in a lateral cluster; stems and leaves triangular; bracts 2-lobed at the summit; stigmas 3 ... ... *pungens*
Spikelets many-flowered, in a terminal umbel or cyme.
    Stems triangular; leaves broad-linear, keeled, but otherwise flat; involucral bracts leafy, the lowest long and erect ... ... *maritimus*
    Stems round, or somewhat compressed towards the summit; leafless except sheathing scales; involucral bracts short, erect, rigid; stigmas 2.
        Hypogynous bristles 6, filiform, with reflexed hairs ... ... *lacustris*
        Hypogynous scales 4 to 6, flattened, plumose with lax hairs ... ... *litoralis*

Isolepis.

Spikelet solitary terminal, stigmas 2, fruit biconvex, stamens 3. Usually floating; stems elongated, with filiform leaves at the nodes ... ... *fluitans*

Spikelet solitary or clustered, terminal. Stigmas 3. Stamens 3; bracts prominently keeled.
    Fruit ovoid-globular, 3-ribbed ... ... *setaceus*
    Fruit obtusely triquetrous, smooth ... ... *riparius*
    Fruit acutely triquetrous, smooth; bract prominently ribbed ... ... *cartilagineus*

Stamen 1, bracts broad scarcely keeled; fruit acutely triquetrous; stems sometimes dwarf, but often elongate and proliferous ... ... *inundatus*

Spikelets clustered, lateral, the outer involucral bract erect and continuous with the stem; stems leafless, robust; stigmas 3.
    Spikelets 2 to 6, sessile, oblong; fruit transversely striate... ... ... *supinus*
    Spikelets numerous, in a dense globular head; fruit smooth ... ... *nodosus*

Lipocarpha.

Dwarf tufted annual, stems very slender, fruit oblong flattened... ... ... *microcephala*

Carex.

I. Spikelets each with male and female flowers, stigmas 2.

Spikelets several in a short terminal spike, bracts pointed greenish.
    Spikelets 3 to 5, male flowers at the base, ovate; outer involucral bract exceeding the inflorescence *inversa*
    Spikelets 6 to 12, male flowers at the top of each *chlorantha*

Spikelets numerous in a long narrow panicle.
    Stems cylindrical; leaves linear, revolute... ... *tereticaulis*
    Stems triangular; leaves broad-linear, very long... *paniculata*

II. Spikelets few or many, the terminal one with male flowers.

Style-branches 2; spikelets 3 to 6, erect, sessile ... *caespitosa*

Style-branches 3, spikelets few, utricles beaked.
    Utricle of corky texture, ovoid, with a short beak, ¼ in. long ... ... *pumila*
    Utricle of thin texture, much smaller ... ... *breviculmis*
    Utricle of thin texture, obtusely 3-angled, tapering into a long beak ... ... *Gunniana*

Style-branches 3; spikelets numerous, cylindrical, on
  long drooping stalks ... ... *pseudocyperus*

Caustis.

Branches robust, flattened; spikelets rather large;
  bracts pubescent ... ... *pentandra*

Cladium.

I. Spikelets when 2-flowered, the lowest fertile,
    its bract as long as the outer empty one.

Panicle densely corymbose; stems very tall, leafy
  throughout; leaves very long, broad-linear, flat,
  the keel and edges rough ... ... *mariscus*
Panicle loose or narrow; leaves chiefly radical, rarely 0.
  Leaves cylindrical; spikelets 2- or 3-flowered.
    Leaves hollow, transversely partitioned; spikelets
      numerous, not clustered; panicle large, some-
      what drooping ... ... *articulatum*
    Leaves solid, or obscurely partitioned; spikelets
      densely clustered; panicle erect; *bracts ciliate*... *glomeratum*
  Leaves angular, or flat with a prominent midrib; spike-
    lets 1-flowered; panicle hardly spreading ... ... *tetraquetrum*
  Leaves vertically flattened, narrow; panicle contracted *schoenoides*
  Leaves very long, cylindrical, channelled, rough; stems
    very tall, leafy; spikelets crowded, in axils of long-
    pointed leafy bracts, forming a long narrow panicle;
    stamens 3; fruit narrow triquetrous ... ... *filum*
Stems leafless except short points to the sheathing
  scales.
    Bract spreading, twice as long as the fruit ... *Gunnii*
    Bract appressed, about as long as the fruit ... *junceum*

  II. Spikelets when 2-flowered, the lowest sterile,
    its bract obtuse longer than the fertile
    one. *(Gahnia)*.

Stems leafy, very tall, or tall; leaves long with rough
  involute margins ending in long subulate points.
    Panicle long and narrow; bracts with rigid erect
      points; stamens 4 to 6; stigmas 3; fruit obovoid-
      oblong, not angled. Resembles *C. filum* ... *trifidum*
    Panicle very compound, with erect branches, black;
      fruit triangular, minutely granular; stamens 3;
      stigmas 3 ... ... *radula*
    Panicle large with spreading or drooping branches;
      fruit ovoid, scarlet; stamens 4 to 6; stigmas 3,
      generally bifid ... ... *psittacorum*

Leaves radical; rather dwarf tufted perennials; panicle narrow erect; spikelets small; leaf-sheaths woolly at orifice.
   Leaves smooth, subulate, long-pointed; spikelets distinct ... ... *lanigerum*
   Leaves with scabrous involute margins ending in long subulate erect points; spikelets clustered... *deustum*

Lepidosperma.

I. Stems hollow, compressed, several feet high.
Leaves blunt-edged, rather flaccid, panicle contracted   *longitudinale*
Leaves rather acute-edged and rigid, panicle spreading   *exaltatum*

II. Stems solid.

Stems broad, tall, flattened but convex on both sides; panicle large, very compound.
   Panicle dense, short; leaves about ½ in. wide, with a broad acute edge. Sandy sea-shores ... ... *gladiatum*
   Panicle elongate and spreading; leaves narrower... *elatius*

Stems narrow, compressed, slightly convex or flat.
   Stems flat, sharp-edged, from 1½ to 3 lines wide.
     Panicle elongated and narrow, *much exceeding* the lowest involucral bract ... ... *laterale*
     Panicle short and somewhat spreading ... *concavum*
   Stems flat, rough-edged, about 2 lines wide, margins resinous; panicle narrow, rather dense ... *viscidum*
   Stems flat, blunt-edged, 1 to 1½ lines wide; panicle spike-like or interrupted.
     Spikelets in sessile clusters, bracts acutely acuminate ... ... *congestum*
     Spikelets in globose clusters, bracts acuminate   *globosum*
   Stems convexly flattened, under 1 line broad.
     Stems almost semicylindrical, panicle spike-like, spikelets narrow and pointed ... ... *semiteres*
     Stems flat; panicle slender but quite short, often exceeded by the lowest involucral bract   *lineare*

Stems filiform or cylindrical.
   Panicle compound, contracted; stems filiform, smooth; leaves somewhat compressed and channelled ... ... *canescens*
   Spikelets scattered in a spike-like panicle; stems filiform-cylindrical; leaves almost undeveloped   *filiforme*
   Spikelets crowded; stems terete grooved on one side, leaves similar but shorter ... ... *carphoides*

Chorizandra.

Stems rigid; leaves few, terete; fruit ovoid, 8-ribbed   *enodis*

GENERA AND SPECIES.

ORDER GRAMINEAE.

1. One fertile flower with or without barren ones in each spikelet.
   A. Pedicel of the spikelet jointed below the glumes. Outer glumes 3

Involucral bristles supporting each spikelet.
   Bristles whorled; spikelets in cylindrical panicles — **Setaria**
   Bristles unilateral; spikelets in a simple panicle ... **Pennisetum**
Outermost glume often minute; spikelets arranged in a spreading panicle, or spike-like; one barren flower in the spikelet ... **Panicum**
Spikelets *unisexual*, arranged in dense heads ... **Spinifex**
Spikelets in a dense spike, 1 or 2 outer *glumes ciliate* ... **Neurachne**
Spikelets arranged in pairs; one spikelet in each pair fertile and sessile, the other usually sterile.
   Spikelets in alternate pairs in the notches of the rhachis of a simple spike ... **Hemarthria**
   Spikelets in a dense cylindrical panicle; glumes concealed under long silky hairs; stamens 1 or 2 **Imperata**
   Both spikelets fertile invested in long soft hairs; spikelets in cylindrical panicles; stamens 3 or 2; two of the *glumes awned* ... **Erianthus**
Fertile spikelets supported by 1, 2, or 4 barren spikelets; one or more glumes awned.
   One or two sterile spikelets supporting the fertile one; inflorescence panicled or spicate ... **Andropogon**
   Four sterile spikelets supporting the fertile one, within sheathing leafy bracts ... **Anthistiria**

   B. Pedicel of the spikelet jointed below the glumes. Outer glumes 2.

Spikelet with a *callous ring* at the base; flowering glume shortly awned; spikelets 1-flowered ... **Eriochloa**
Spikelets 1-flowered, in a loose narrow spike or raceme; the *outer glumes with long straight awns* ... **Perotis**
Spikelets 1-flowered, not awned; the larger *outer glume with short hooked bristles*; inflorescence spike-like **Tragus**

   C. Pedicel of the spikelet jointed above the glumes; outer glumes 3.

Spikelets 1-flowered, *2 additional bracts* below the articulation, two of the glumes awned; stamens 4; spikelets in panicles ... **Ehrharta**
Spikelets 2- or 3-flowered; flowering glumes with 9 plumose awns; inflorescence a spike-like panicle... **Pappophorum**
Spikelets 1-flowered in the alternate notches of the rhachis of a simple spike. (Resembles *Hemarthria*) **Lepturus**

D. Pedicel of the spikelet jointed above the glumes; outer glumes 2; spikelets 1-flowered.

Flowering glume awned.
  Awn simple and terminal.
    Glume 3-lobed, central lobe awned; rhachis produced in a small bristle ... ... **Echinopogon**
    Glume keeled, awn very short, spiklelets crowded in a cylindrical spike-like panicle **Alopecurus**
    Glume rounded, rolled around the flower; awn very long, spirally twisted; spikelets in branched panicles, rarely spike-like; lodicules large ... ... **Stipa**
    Glume 3-lobed or entire; spikelets singly sessile in 2 rows on one side of simple spikes digitate at the end of the peduncle ... ... **Chloris**
  Awn simple, not terminal, dorsal or basal.
    Glume on a short hairy stalk, membranous; awn from a little below the end, scarcely twisted; panicle spike-like ... ... **Dichelachne**
    Awn nearly basal, or about the middle, usually twisted; panicle loose and spreading, or spike-like ... ... **Agrostis**
  Awn 3-branched terminal; leaves subulate ... **Aristida**
  Awns 3; flowering glume on a short hairy stalk; panicle spike-like, cylindrical or oblong. (In some species of *Chloris* the 2 outer lobes of the flowering glume are shortly awned ... ... **Amphipogon**
  Awns 5, 1 long and 4 small; spikelets in panicles **Pentapogon**

Flowering glume awnless (also *Agrostis* partly).
  Spikelets in a loose or narrow panicle ... ... **Sporobolus**
  Spikelets in 2 rows on one side of simple spikes, the spikes digitately grouped at the end of the peduncle ... ... **Cynodon**

      II. Two perfect flowers in each spikelet.

Flowering glume *truncate 4-toothed;* awn dorsal; outer glumes transparent; panicle much spreading ... **Aira**
Flowering glume hairy; awn terminal short or 0; outer glumes many-nerved, acute or shortly awned **Eriachne**

      III. Three or more perfect flowers in each spikelet.

        a. Spikelets awned.

Flowering glume 3-lobed, 3-awned; spikelets in panicles **Triraphis**
Flowering glume 3-nerved tapering into short awns; *stamen* 1; *spikelets* in *globular* or *cylindrical spikes* **Elytrophorus**

Flowering glume several-nerved; awn terminal, between rigid lobes or lateral awns, rarely infra-terminal; spikelets in panicles ... ... **Danthonia**

Flowering glume 3-lobed, central one awned; *spikelets sessile in 2 rows on one side* of 1 or 2 simple spikes **Astrebla**

Glumes awned; *spikelets sessile, alternate* on the sides of a simple spike, their flat side turned to the rhachis ... ... **Agropyron**

Glumes with long awn-like points; rhachis with long hairs enveloping the flowers; panicle large, dense **Arundo**

Grain adnate to the palea, flowering glumes awned.
    Ovary pubescent; flowering glume with a hyaline tip, the awn attached below it; panicle small ... **Bromus**
    Ovary glabrous, flowering glume acute, awn terminal; panicle loose ... ... **Festuca**

    b. Flowering glume only minutely pointed.

Flowering glume with 2 hyaline lobes besides the terminal point; inflorescence spicate or in narrow panicles ... ... **Diplachne**

Flowering glume obtuse or notched, 5-nerved, the central nerve minutely pointed; panicle narrow, long **Schedonorus**

    c. Spikelets awnless.

Spikelets arranged in spikes.
    Spikelets unisexual on separate plants ... ... **Distichlis**
    Spikelets digitate at the end of the peduncles ... **Eleusine**

Spikelets in panicles.
    Flowering glume 3-lobed or 3-toothed; leaves pungent-pointed ... ... **Triodia**
    Flowering glume lobeless; spikelets compressed.
        Flowering glume 5-nerved; spikelets many-flowered ... ... **Eragrostis**
        Flowering glume 3-nerved; spikelets usually few-flowered ... ... **Poa**

Setaria.

Flowering glume rugose; bristles scabrous with erect teeth.
    Panicle cylindrical, short; spikelets solitary at the base of the bristles. Pale-green annual ... *glauca*
    Panicle dense or interrupted, about 6 in.; spikelets clustered near the base of the bristles. Taller and stouter ... ... *macrostachya*

Flowering glume smooth; panicle loosely cylindrical, about 2 in.; bristles scabrous with erect teeth ... *viridis*

Pennisetum.

Much branched, glabrous annual; bristles not plumose — *refractum*

Panicum.

a. Lower branches of the panicle whorled, upper ones scattered.

Spikelets silky hairy, in pairs along one side of the branches.
    Uppermost glume 5- to 7-veined; spikelets 1½ to 2 lines long ... *coenicolum*
    Upper glume 3-veined; spikelets 1 to 1½ lines ... *divaricatissimum*
Spikelets glabrous, scattered; uppermost glume 5- to 7-veined; ligule very prominent, not ciliate ... *prolutum*

b. Lower branches of the panicle clustered, upper ones scattered.

Lowest glume acute, half as long as the spikelet.
    Nodes prominently ciliated, leaves hairy ... *effusum*
    Nodes and leaves glabrous, ligule very short ... *Mitchelli*
Lowest glume truncate, very short. Tall, glabrous; ligule very short, ciliate ... *decompositum*

c. Branches of the panicle scattered, spreading (also *P. distachyum* and *P. reversum*).

Branches ending in awn-like points; spikelets distant ... *spinescens*
Spikelets sessile, crowded.
    Spikelets *intermixed with bristles;* uppermost glume often *awned;* fruiting glume smooth ... *Crus-galli*
    Fruiting glume rugose, tipped with a minute point ... *adspersum*
Spikelets stalked; fruiting glume smooth, stalked. Much branched, beset with long hairs ... *pauciflorum*

d. Panicle spike-like, simple or of a few erect branches.

Spikelets beset with long silky hairs; flowering glume smooth.
    Spikelets rather acute; innermost glume 5-nerved ... *leucophaeum*
    Spikelets truncate; innermost glume nerveless ... *argenteum*
Spikelets close together in 2 rows.
    Spikelets glabrous; flowering glume acute wrinkled ... *gracile*
    Spikelets somewhat hairy; flowering glume obtuse, with an awn-like point; leaves hairy.
        Innermost glume 5-nerved, flowering glume minutely rugose ... *helopus*
        Innermost glume 3- to 5-nerved, ciliate on the margin ... *Gilesii*

GENERA AND SPECIES. 191

Spikes at first erect, at length spreading or reflexed; spikelets alternate along the rhachis.
    Rhachis slender or slightly dilated, slightly hairy … *distachyum*
    Rhachis flat, ending in an awn-like point; a rigid bristle under the lowest spikelet … … … *reversum*

Spinifex.
Glabrous, erect; branches clustered surrounded by short leaves … … … … … … *paradoxus*
Silky-pubescent leaves; branches robust, extensively creeping in sand by the sea. Heads of spikelets several inches diameter … … … … *hirsutus*

Neurachne.
Spike ovoid, about 1 in.; outer glume 5- or 7-nerved, with long spreading hairs on the back. Erect glabrous … … … … … … … *alopecuroides*
Spike narrow, 1 to 2 in. long. Stems from a woolly base.
    Outer glume with a transverse callosity bearing long cilia … … … … … … *Mitchelliana*
    Outer glume thin, glabrous or bordered by a few cilia … … … … … … *Munroi*

Hemarthria.
Slightly branched, ascending to 1 foot; spikelets closely appressed … … … … *compressa*

Imperata.
Tall, stiff, erect, glabrous; leaves erect often longer than the stem … … … … … … *arundinacea*

Erianthus.
Stems slender, sometimes tall; silky hairs of spike rich-brown … … … … … … *fulvus*

Andropogon.
Spikes 2 or more clustered at the end of the peduncle.
    Spikelets concealed under copious silky hairs.
        Outer glumes, rhachis and stalklets hairy. *Nodes bearded* … … … … *sericeus*
        Rhachis and stalklets only or chiefly hairy; outer *glumes* marked *with a pit* on the back … *pertusus*
    Spikelets silky hairy, but not concealed.
        Spikes 3 or 4, quite terminal, in a close cluster *annulatus*
        Spikes many, the common axis elongated … *punctatus*

Spikes sessile, 2 together, within a sheathing bract,
at the end of each peduncle, forming a contracted
panicle; spikelets concealed or nearly so by long
silky hairs.
    Spikes erect; awns prominent ... ... *exaltatus*
    Spikes reflexed; awns none or very short ... ... *bombycinus*
Spikes in elongated panicles, the branches whorled,
glabrous; outer glume with a long twisted awn ... *Gryllus*

Anthistiria.

Barren spikelets sessile. Awns rigid.
    Fertile spikelet glabrous; clusters of spikelets
sessile ... ... *ciliata*
    Fertile spikelet densely silky; clusters of spikelets
stalked ... ... *avenacea*
Barren spikelets stalked. Awns very fine ... ... *membranacea*

Eriochloa.

Rhachis of the spike hairy or glabrous, spikelets about
1 in. long ... ... *polystachya*

Perotis.

Slender ascending to one foot; leaves linear, slender-
pointed ... ... *rara*

Tragus.

Spreading annual; leaves flat, margins ciliated ... *racemosus*

Ehrharta.

Stems slender erect, 1 to 2 ft.; leaves short revolute,
glabrous ... ... *stipoides*

Pappophorum.

Stems erect about 1 ft.; outer glumes beset with soft
hairs ... ... *commune*

Lepturus.

Spike often curved; lowest glumes 2; stems stiff,
dwarf; leaves short, quite narrow, incurved along
the margin ... ... *incurvatus*
Spike often straight; lowest glume 1; leaves rather
narrow ... ... *cylindricus*

Echinopogon.

Spikelets in ovoid-globular heads; stems erect 1 or 2 ft.;
leaves flat very scabrous ... ... *ovatus*

Alopecurus.

Stems prostrate in the lower portion, thence abruptly bent upwards. Annual, glabrous; leaves lax, flat — *geniculatus*

Stipa.

I. Flowering glume glabrous. Panicle-branches hairy, lower ones whorled.

Panicle-branches with long hairs, spikelets 4 to 6 inches long ... ... *elegantissima*

Panicle-branches with short hairs; spikelets smaller; outer glumes short ... ... *Tuckeri*

II. Flowering glume hairy, its hyaline margin dilated on each side of the awn; palea as long as the glume.

Panicle narrow and compact.
Outer glumes colourless; leaves very long, cylindrical, pungent ... ... *teretifolia*
Outer glumes acute, yellowish; leaves flat or convolute; ligule short, not ciliated ... ... *flavescens*

Panicle of 1 to 3 flowers; leafless; stems branched, rampant extending for a few feet ... ... *Muelleri*

III. Flowering glume hairy, its margin not dilated; palea not so long as the glume.

Ligule elongated, not ciliated; panicle loose; leaves slender, filiform; lowest glume fine-pointed; awn slightly rough ... ... *setacea*

Ligule short, ciliate; awn plumose-hairy in the lower part; panicle dense ... ... *semibarbata*

Ligule short, ciliate; awn glabrous or slightly pubescent.
Lowest glume usually dilated and truncate or toothed; flowering glume narrow; panicle dense — *pubescens*
Lowest glume usually 3-pointed; flowering glume rather broad; panicle very loose ... ... *aristiglumis*
Lowest glume always fine-pointed; panicle loose; leaves slender, glabrous or pubescent ... ... *scabra*

Chloris.

Spikes slender, about 10 in number, about 3 in. long; spikelets acute.
Flowering glume bifid, awnless, usually scabrous — *pectinata*
Flowering glume awned or very minutely toothed — *acicularis*

Spikes slender, 6 to 10, 3 to 6 in. long; spikelets cuneate, truncate; flowering glume, obtuse ... ... *truncata*

Spikes dense, 1 to 2 inches long.
   Flowering glume membranous, rather acute; spikes
     6 to 10... ... ... *barbata*
   Flowering glume broad, rigidly scarious, ciliate ... *scariosa*

Dichelachne.

Panicle very dense, almost spike-like; spikelets very
  numerous small concealed by the long hair-like
  awns; flowering glume scabrous. Tall robust ... *crinita*
Panicle rather loose and narrow; awns shorter; flowering glume minutely pitted. Stems slender ... *sciurea*

Agrostis.

    I. Palea very short or none; panicle spreading
Awnless; palea 0 or very minute; leaves tufted narrow   *scabra*
Awn nearly basal; leaves finer; outer glumes longer ... *venusta*

    II. Palea more than half as long as the glume.

Flowering glume about as long as the outer ones, acute,
  often minutely scabrous; awn from about the middle; rhachis of the spikelet produced into a hairy
  bristle. Panicle spike-like ... ... *densa*
Flowering glume much shorter than the others.
  Panicle loose and spreading, rhachis of spikelets
    produced into a hairy bristle; awn somewhat
    basal ... ... *Solandri*
  Panicle spike-like; awn almost basal.
    Rhachis-bristle minute or wanting. Panicle
     short or to 10 in. long... ... ... *quadriseta*
    Rhachis-bristle conspicuous, hairy; panicle 2
     to 4 in. long ... ... *montana*

Aristida.

Awn 3-branched far above the base, the basal part
  spirally twisted, articulate on the glume. Flower-
  ing glume short.
  Awn 1½ in. below the branches; branches about
    2 in. long ... ... *stipoides*
  Awn shorter ; branches usually longer ... ... *arenaria*
Awn 3-branched from its base, not articulate on the
  flowering glume, which is as long as the outer ones.
  Panicle-branches very long, with few spikelets on
    long thin pedicels; outer glumes unequal, with
    long points ... ... *leptopoda*
  Panicle short, broad, and dense; outer glumes
    nearly equal, the 2nd rather longer than the
    flowering glume about ½ inch ... ... *Behriana*

Panicle narrow, rather loose; outer glumes as long
as the flowering one.
    Outer glumes scarcely 3 lines. Awns under
    ½ in. long ... ... *ramosa*
    Outer glumes 4 to 5 lines. Awns ¾ to 1 in.
    long ... ... *calycina*

Amphipogon.

Stems not tall from a creeping rhizome; leaves erect,
rather short, subulate, glabrous. Outer glumes
entire, rather acute ... ... *strictus*

Pentapogon.

Erect, somewhat tall; leaves narrow, pubescent; panicle
narrow, not long ... ... *Billardieri*

Sporobolus.

I. Panicle spike-like.

Dwarf and usually prostrate; leaves short, rigid, somewhat 2-seriate; outer and flowering glumes nearly
equal ... ... *Virginicus*
Erect, rather tall; leaves rather long; outer glumes
unequal, shorter than the flowering one ... ... *Indicus*

II. Panicle spreading, lower branches whorled.

Spikelets stalked, about ½ line long; glumes very acute,
dark coloured ... ... *Lindleyi*
Spikelets sessile, crowded, about 1 line long; outer
glume hyaline obtuse ... ... *actinocladus*

Cynodon.

Flowering glume longer than the outer ones. Prostrate,
rooting at the nodes; stems erect; spikes 2 to 5,
very narrow, dark-coloured ... ... *Dactylon*
Flowering glume much shorter than the outer ones.
    Flowering glume hairy on the keel and margins;
    palea with 2 converging nerves ... ... *convergens*
    Flowering glume with a ring of hairs below the tip;
    palea with 2 distant nerves ... ... *ciliaris*

Aira.

Stems 2 to 4 feet high; leaves stiff, narrow, rough above *caespitosa*

Eriachne.

I. Awn not longer than the glumes.

Panicle loose; leaves glabrous flat; outer glumes hairy *aristidea*

II. Awn absent or reduced to a very small point.
Panicle dense, ovate or oblong; leaves very narrow, long-pointed ... *ovata*
Panicle narrow. Rather tall, slender, with glabrous flat leaves ... *pallida*
Panicle loose or reduced to 2 or 3 spikelets.
    Leaves ½ to ¾ in. long, spreading, pungent-pointed *scleranthoides*
    Leaves not pungent, the upper ones distant.
        Flowering glumes tipped with short points ... *mucronata*
        Flowering glumes obtuse or scarcely acute ... *obtusa*

Triraphis.

Glabrous, about 2 feet high; panicle soft and dense ... *mollis*

Danthonia.

I. Flowering glumes 2-lobed, more or less hairy.
Flowering glumes cleft to near the base, lobes lanceolate, hairs arranged in 2 transverse rows ... *bipartita*
Flowering glumes cleft to near the middle, hairs clustered *carphoides*
Flowering glumes cleft to less than the middle, awn longer than the lobes, hair-tufts in 2 transverse rows ... *penicillata*
    II. Flowering glumes not cleft, minutely denticulated at the summit, without any tufts of hairs; awn infra-terminal ... *nervosa*

Astrebla.

Spikelets closely imbricate; awn about as long as the lateral lobes ... *pectinata*
Spikelets distant, almost erect; awn longer than the lobes ... *triticoides*

Agropyron.

Spikelets narrow, erect, and distant, with long awns; stems rough ... *scabrum*

Elytrophorus.

Erect glabrous dwarf annual; leaves flat; spikes very short and broad ... *articulatus*

Arundo.

Very tall, short stems and long leaves from a creeping rhizome. Semiaquatic ... *Phragmites*

Bromus.

Rather dwarf, annual; leaves flat, flaccid, softly hairy *arenarius*

Festuca.

Erect, rather tall; spikelets ½ in. or more; awns as
long as the glumes ... ... *duriuscula*

Diplachne.

Spikes slender, simple, 2 to 4 in. long, on a long
peduncle... ... ... *loliiformis*
Spikes numerous in a simple panicle.
   Spikelets pale-coloured; rhachis hair-tufted under
   the glumes ... ... *Muelleri*
   Spikelets dark-coloured; rhachis glabrous or
   nearly so ... ... *fusca*

Schedonorus.

Tall; leaves cylindrical, erect, rigid, pungent-pointed,
glabrous, pale yellow; panicle narrow, dense and
spikelike... ... ... *litoralis*

Distichlis.

Prostrate in broad patches; leaves short, pungent-
pointed, usually spreading in 2 rows; flowering
stems leafy ... ... *maritima*

Eleusine.

Spikes digitate, usually 4; spikes closely packed.
Dwarf annual ... ... *cruciata*
Spikes 6 to 12, mostly crowded at the end of the
rhachis ... ... *digitata*

Triodia.

Flowering glume divided, nearly to the middle, into
3 lobes, silky hairy; leaf-sheaths usually viscid.
   Panicle loose and spreading; spikelets dark-coloured, 8- to 12-fld. .. .. *Mitchelli*
   Panicle narrow and dense; spikelets pale-coloured,
   about 6-fld. ... ... *pungens*
Flowering glume shortly 3-toothed, silky-hairy at the
base; leaf-sheaths not viscid... ... ... *irritans*

Eragrostis.

1. Spikelets rather flat; glumes rather distant.
Spikelets 3- to 4-flowered, very numerous and minute,
stalked, in a spreading panicle ... ... *tenella*
Spikelets linear, more than 6-flowered, numerous, in a
loose panicle.
   Glumes very obtuse, truncate or emarginate ... *trichophylla*

Glumes acute.
 Spikelets crowded on the long branches of a narrow panicle ... ... *leptocarpa*
 Spikelets distant; panicle-branches erect, capillary ... ... *pilosa*

 II. Spikelets very flat; glumes closely imbricate.

Base of the stems glabrous, not at all or scarcely thickened.
 Spikelets in small globose or oblong clusters, sessile along an unbranched rhachis. Stamens usually 2 *diandra*
 Spikelets erect, scattered or clustered in a simple or branched panicle. Stamens usually 3 ... *Brownii*
 Spikelets broad, crowded on a short almost simple rhachis; nerves of the palea with long rigid hairs *concinna*
 Spikelets narrow, clustered along the erect branches of a narrow panicle; palea truncate, glabrous ... *speciosa*

Base of the stems and short sheath of radical leaves thickened into a bulbous woolly-hairy base.
 Spikelets shortly stalked, nearly 2 lines broad; base of the flowering glumes woolly ... ... *laniflora*
 Spikelets about ¾ line broad, glabrous ... ... *chaetophylla*
 Spikelets sessile, above 1 line broad, glabrous ... *eriopoda*

 III. Spikelets terete or nearly so, very narrow; glumes closely appressed.

Spikelets short-stalked, in a small panicle, rather obtuse, 10- to 30-flowered ... ... *lacunaria*
Spikelets sessile, obtuse, 12- to 50-flowered, usually clustered, often incurved ... ... *falcata*

Poa.

 I. Lodicules disunited. Stigmatic plumes not branched; flower-glumes keeled.

Perennial; grain adnate to the palea; panicle narrow and dense, the spikelets crowded. Tall, rigid, coast-grass ... ... *Billardieri*

Perennials; grain free from the palea.
 Stems knotty at the base; leaves flat; panicle short ... ... *nodosa*
 Panicle dense and contracted or spreading; leaves flat, longitudinally incurved, ending in long points ... ... *caespitosa*

Annual, leaves flat flaccid; spikelets compressed; flowering glume 7- to 11-nerved, the keel ciliate at the base with long hairs .. .. *lepida*

GENERA AND SPECIES.

11. Lodicules united. Stigmatic plumes branched; flowering glumes rounded on the back.

Semiaquatic, floating or creeping; panicle narrow, long and loose; flowering glumes glabrous, 7-nerved ... *fluitans*

Erect or diffuse branchless stems.
   Flowering glumes hair-tufted, 7- to 9-nerved; panicle loose ... ... *Fordeana*
   Flowering glumes glabrous, 5-nerved; panicle narrow, dense ... ... *syrtica*

Very tall, branched stems; panicle very spreading; flowering glumes broad, concave, hyaline, 3-nerved *ramigera*

CLASS III.—VASCULAR ACOTYLEDONS.

ORDER LYCOPODIACEAE.

Spore-cases and spores all similar. Comparatively large **Lycopodium**
Spore-cases and spores of two kinds. Small erect plants **Selaginella**

Lycopodium.

Stems creeping; spikes single on lateral erect peduncles *Carolinianum*
Stems erect, branched at the base; spikes sessile, lateral *laterale*
Stems much branched, erect; spikes sessile, terminal ... *densum*

Selaginella.

Stems from a branching base, simple, about 1 inch ... *Preissiana*
Stems branching upwards, several inches long ... ... *uliginosa*

ORDER RHIZOSPERMAE.

Floating minute plants, much branched, reddish; leaves crowded; fruit-masses sessile, axillary, transparent, globular ... ... **Azolla**
Fronds erect, stalked, divided into 4 segments; spore-cases hard, compressed, basal. Semiaquatic; creeping ... **Marsilea**
Fronds linear, erect; spore-cases globular, basal, sessile. Semiaquatic with creeping rhizome ... ... **Pilularia**

Azolla.

Stems once or twice pinnate, broadly ovate in outline, with linear leafy branches, the segments slightly distant; roots feathery ... ... *pinnata*
Segments of the stems short, often closely imbricate; roots simple ... ... *filiculoides*

Marsilea.

Barren fronds glabrous or hairy; leaf-like segments cuneate-ovate ... *quadrifolia*

Pilularia.

Barren fronds 1 to 3 in., bright-green; fruit-masses like little pills, 1½ to 2 lines diameter, slightly hairy, bursting by 4 valves ... *globulifera*

ORDER FILICES.

I. Spore-cases globular, 2-valved, without any ring, sessile in 2 rows. Fertile fronds spike-like, simple or branched, often connate at the base with the barren frond.

Barren frond solitary, undivided; fertile frond, a simple spike ... **Ophioglossum**

Barren frond solitary, segmented; fertile frond, a paniculate spike ... **Botrychium**

II. Spore-cases globular, without any perfect ring, 2-valved; sessile in 2 rows covering the inner surface of the pinnules. Erect, simple, or dichotomous fronds, without expanded laminae ... **Schizaea**

III. Spore-cases globular, with a transverse ring, opening vertically into 2 valves.

Sori of 2, 3, or few spore-cases to each segment of the pinnules. Frond dichotomous, segments of the pinnules in two rows ... **Gleichenia**

Sori of numerous spore-cases on the lower side of much contracted frond-segments. Fronds bipinnate ... **Osmunda**

IV. Spore-cases with a longitudinal ring, ruptured irregularly, stalked; sori on the underside or rarely at the margin of the frond.

    a. Sori covered at least when young with an indusium.

Sori globular, close to the margin; indusium adnate on the upper side, opening in 2 valves ... **Dicksonia**

Sori linear, marginal; indusium membranous, opening from the margin inwards ... **Lindsaea**

Sori marginal; indusium continuous with the margin and opening from the outer edge outwards.
  Sori short; frond compound, veins of the pinnules forked radiating from the stalklet ... **Adiantum**

Sori short or globular, the slightly altered margin bent over them	**Cheilanthes**
Sori and indusium linear, usually long and continuous; fronds compound	**Pteris**
Sori in a continuous line on both sides of the midrib, the indusium opening from the midrib outwards; sori at length covering the underside of the fertile fronds	**Lomaria**
Sori and indusium oblong, or linear, on veins diverging from the midrib...	**Asplenium**
Sori orbicular, usually small; indusium attached within the sorus, peltate or orbicular-reniform	**Aspidium**

b. Sori without indusium.

Sori orbicular, usually small	**Polypodium**
Sori linear or oblong, on veins diverging from the midrib	**Grammitis**
Sori short often confluent; covered by the recurved margin of the frond, forming a spurious indusium...	**Cheilanthes**

Ophioglossum.

Barren frond ovate-lanceolate, sessile near the middle of stem, but distant from the spike. Quite dwarf	*vulgatum*

Botrychium.

Sterile frond long-stalked, ternately and pinnately divided	*ternatum*

Schizaea.

Stalk of the fertile fronds undivided, filiform, channelled	*fistulosa*
Stalk of the fertile fronds, mostly twice divided, soriferous; pinnules linear, flat, about 3 to 4 lines long	*bifida*

Gleichenia.

Pinnules divided to the midrib into many, flat, rather stiff, segments, each with a single sorus	*circinata*

Osmunda.

Rhizome erect forming a short broad trunk; fronds attaining to 6 feet long, glabrous; pinnules of a firm consistence	*barbara*

Dicksonia.

Rhizome arborescent. (Here probably extinct). ...	*Billardieri*

Lindsaea.

Fronds pinnate, rhachis black wiry; pinnules small, distant, obliquely flabellate	*linearis*

Adiantum.

Fronds tripinnate; pinnules broadly ovate, cuneate at at the base, broadly crenate, the sori in the sinus of the crenatures ... ... *Æthiopicum*

Pteris.

Fronds rigid, somewhat hairy below; veins of pinnules diverging ... ... *aquilina*
Fronds flaccid, glabrous, very ample.
    Segments of pinnules narrow-lobed; veins diverging *arguta*
    Segments of pinnules broad-lobed; veins imperfectly reticulate ... ... *incisa*

Lomaria.

Barren fronds with numerous segments, attached to the rhachis by a broad base.
    Barren segments narrow, long; rhachis dark ... *discolor*
    Barren segments lanceolate, short; rhachis pale ... *lanceolata*
Barren segments attached by the midrib only, obliquely truncated at the base ... ... *Capensis*

Asplenium.

Sori linear.
    Frond pinnate, segments quite short denticulate; *rhachis* filiform, very long and *extended* beyond the segments ... ... *flabellifolium*
    Frond bipinnate, the segments prominently veined *furcatum*
Sori oblong, quite dorsal; fronds bipinnate, often developing near the summit small bulbs, which originate new plants ... ... *bulbiferum*

Aspidium.

Fronds pinnate, softly hairy; segments elongate, pinnatifid ... ... *molle*
Fronds repeatedly pinnate, glabrous, with acutely toothed or lobed segments ... ... *decompositum*

Polypodium.

Frond 2- or 3-pinnate, long and wide, beset with short glandular hairs ... ... *punctatum*

Grammitis.

Perennial, tufty; frond beset with scales underneath.
    Frond pinnate; pinnæ broadly ovate, entire; sori concealed by the scales ... ... *Reynoldsii*

Frond pinnate ; segments ovate-cuneate, often oblique, usually incised ; sori often confluent in large patches...	*rutaefolia*
Annual, minute ; frond often solitary, 1- or 2-pinnate, glabrous, thin and delicate ; segments broadly obovate or fan-shaped, each with a single sorus ...	*leptophylla*

Cheilanthes.

Frond glabrous, the recurved margin over the sori slightly altered; 2- or 3-pinnate, ultimate segments small, irregularly crisped at the margin	*tenuifolia*
Frond glabrous, bipinnate ; the indusium as in *Pteris* with the sori as in *Cheilanthes*	*Clelandi*
Fronds beset underneath with a dense investiture, the recurved margin over the sori unaltered.	
Pinnæ densely covered with brown woolly hairs ...	*vellea*
Pinnæ densely covered with colourless bristly scales	*distans*

A CLASSIFIED LIST OF THE NATIVE SPECIES

WITH ANNOTATIONS INDICATING THEIR DISTRIBUTION WITHIN THE PROVINCE.

To record localities for each species would add much to the bulk and cost of this work, but as it seemed desirable to give some idea of their distribution, I have adopted the plan of subdividing the Province into 12 districts (as set forth in the following schedule and the accompanying map), and by the use of monograms to indicate their occurrences therein.

Two chief floras are recognised:—The Eremian or Desert Flora which occupies the arid region of Central Australia and corresponds with the "salt-bush country" of the pastoralist. The region is approximately limited by the rain-fall line of ten inches. 2. The Euronotian Flora which is dominant in the more humid parts of temperate Australia, excepting the extreme south-west.

EREMIAN REGION.

F. North of the Central District, chiefly comprising the basin of the upper Finke-river and its tributaries.

C. Central District. This comprises chiefly the low plains around Lake Eyre and is demarked by the rain-fall line of 7 inches in conjunction with certain physical features. Its flora is most characteristically eremian.

S. South of District C, extending from Lake Torrens to the Barrier Range; it overlaps N and M.

W. West of Lake Torrens, overlapping C on the north and L on the south.

M. The plain of the Lower Murray River. It is defined on the west by the Adelaide chain and its north-east extension to the Barrier Range.

EURONOTIAN REGION.

A. The Adelaide District.

N. The northern agricultural areas, separated from A by a line drawn from the head of St. Vincent-gulf to Burra.

Y. Yorke-Peninsula.

L. The Port Lincoln district, comprising southern Eyre-peninsula and the costal tract extending towards the Head of the Great Bight.

K. Kangaroo Island.

T. South of the Murray Desert embracing the "90-mile desert" and the Tatiara.

G. The volcanic area of the south-east corner of the Province or the Mount Gambier district.

LIST OF SPECIES.

RANUNCULACEAE.
Clematis, Linne (1737).

aristata, *R. Brown*									G
microphylla, *DeCandolle*	S	.	M	A	N	Y	L	K	T	G

Ranunculus, Linne (1737).

aquatilis, *Linne*	S	.	M	A	.	.	L	.	T	G
lappaceus, *Smith*	S	.	.	A	N	.	L	.	T	G
rivularis, *Banks & Solander*	.	.	M	A	N	.	.	K	T	G
parviflorus, *Linne*	C S W	M	A	N	Y	L	K	T	G

Myosurus, Linne.

minimus, *Linne*	S	.	M	.	.	.	.	T	

DILLENIACEAE.
Hibbertia, Andrews (1800).

hirsuta, *Bentham*				A					
sericea, *Bentham*				A	.	.	L	K	T	G
stricta, *R. Brown*			M	A	N	Y	L	K	T	G
Billardieri, *F. v. Mueller*				A	.	.	.	K		G
acicularis, *F. v. Mueller*				A	.	.	.	.	.	G
virgata, *R. Brown*		W	M	A	.	Y	.	K	.	.
fasciculata, *R. Brown*				A	.	.	L	K	T	G
glaberrima, *F. v. Mueller*	F									

LAURACEAE.
Cassytha, Osbeck (1753).

glabella, *R. Brown*			A	.	Y	L	K	.	G
pubescens, *R. Brown*		M	A	.	Y	L	K	.	G
melantha, *R. Brown*	W	M	A	.	Y	L	K	T	G

CERATOPHYLLEAE.
Ceratophyllum, Linne (1735).

demersum, *Linne*	M

PAPAVERACEAE.
Papaver, Linne.

aculeatum, *Thunberg*	M	A	N	Y	L	K	.	.

CAPPARIDAE.
Cleome, Linne.

viscosa, *Linne*	F

Capparis, Linne.

```
lasiantha, R. Brown     ...  ...   C
spinosa, Linne ...      ...  ...   F
Mitchelli, Lindley      ...  ...   F C S . M
```

CRUCIFERAE.

Nasturtium, R. Brown (1812).

```
terrestre, R. Brown     ...  ...  ...  ...   M A             T G
```

Cardamine, Linne.

```
eustylis, F. v. Mueller ...  ...  ...  ...   M
laciniata, F. v. Mueller     ...  ...  ...   M A . Y         T G
flexuosa, Withering     ...  ...  ...  ...  A N                G
```

Barbarea, Beckmann (1801).

```
vulgaris, R. Brown      ...  ...  ...  ...  A
```

Erysimum, Linne.

```
curvipes, F. v. Mueller ...  ...   S W M . N
brevipes, F. v. Mueller ...     C  S W M . N Y
lasiocarpum, F. v. Mueller  ... F C . . M . . Y    T .
Blennodia, F. v. Mueller    ... F C S W
```

Sisymbrium, Linne.

```
filifolium, F. v. Mueller    ...    S W . N
trisectum F. v. Mueller      ... F C S W M
nasturtioides, F. v. Mueller ...   S W M . N
procumbens, Tate        ...  ...   S
Richardsii, F. v. Mueller    ...      W
cardaminoides, F. v. Mueller ... C        M        T G
```

Cakile, Linne.

```
maritima Scopoli    ...  ...  ...  ...  ...  ...  ... K
```

Stenopetalum, R. Brown (1821).

```
velutinum, F. v. Mueller     ... F C . . M
lineare, R. Brown       ...  ... F . S . M A N Y L    . G
nutans, F. v. Mueller   ...  ... F C
sphaerocarpum, F. v. Mueller ...  ...  ... A N Y       T
trisectum, Tate ...     ...  ... C
```

Geococcus, Drummond & Harvey (1855).

```
pusillus, Drummond & Harvey   ...    W M        Y
```

Alyssum, Linne.

```
minimum, Pallas    ...  ...  ...   S W M . N Y
```

LIST OF SPECIES.

Menkea, Lehmann (1843).

australis, *Lehmann* ... ... M
sphaerocarpa, *F. v. Mueller* ... F

Capsella, Moench (1792).

pilosula, *F. v. Mueller* ... ... W M   Y
elliptica, *F. v. Mueller*... ... ... S W M A N Y L K T G
humistrata, *F. v. Mueller* ... ... M
cochlearina, *F. v. Mueller* ... ... S W   N
Drummondi, *F. v. Mueller* ... ... W

Lepidium, Linne.

strongylophyllum, *F. v. Mueller*   F
leptopetalum, *F. v. Mueller* ... ... S   M
rotundum, *DeCandolle*... ... F C
phlebopetalum, *F. v. Mueller*... F C S   M
monoplocoides, *F. v. Mueller* ...   C   M   Y
papillosum, *F. v. Mueller* ... F C S W M   N
ruderale, *Linne* ... ... F C S . M A N Y . K T G
foliosum, *Desvaux* ... ... K T

VIOLACEAE.

Viola, Linne.

hederacea, *Labillardiere* ... ... A   K   G
betonicifolia, *Smith* ... ... A   G

Hybanthus, Jacquin (1763).

floribundus, *F. v. Mueller* ... ... W   A   T G
enneaspermus, *F. v. Mueller* ... F
Tatei, *F. v. Mueller* ... ... S

Hymenanthera, R. Brown (1818).

Banksii, *F. v. Mueller* ... ... A

DROSERACEAE.

Drosera, Linne (1737).

binata, *Labillardiere* ... ... A . . . . . G
glanduligera, *Lehm.* ... ... A   Y L K T G
Whittakerii, *Planchon*... ... ... A   Y . K T G
pygmaea, *DeCandolle* ... ... A   L K T G
spathulata, *Labillardiere* ... ... G
Burmanni, *Vahl*... ... ... F
Indica, *Linne* ... ... F
Menziesii, *R. Brown* C   A N Y L K . G
auriculata, *Backhouse* ... ... S   A N   K . G
peltata, *Smith* ... ... A N . L   T G

FRANKENIACEAE.
Frankenia, Linne.
laevis, *Linne* ... ... F C S W M A N Y L K T G

PITTOSPOREAE.
Pittosporum, Banks (1788).
phillyraeoides, *DeCandolle* ... F C S W M A N Y L K T G

Bursaria, Cavanilles (1797).
spinosa, *Cavanilles* ... . S . M A N Y L K T G

Marianthus, Huegel (1837).
bignoniaceus, *F. v. Mueller* ... ... A   K

Billardiera, Smith (1793).
scandens, *Smith* ... ... L K   G
cymosa, *F. v. Mueller* ... ... A N Y L K T G

Cheiranthera, Cunningham (1829).
linearis, *Cunningham* ... ... A N Y L K
volubilis, *F. v. Mueller* ... ... K

POLYGALEAE.
Polygala, Linne.
Chinensis, *Linne* ... ... F

Comesperma, Labillardiere (1806).
scoparium, *Steetz* ... ... W M   L
volubile, *Labillardiere* ... ... A   Y L K . G
sylvestre, *Lindley* ... ... F . S
viscidulum, *F. v. Mueller* ... F
calymega, *Labillardiere* ... ... A   L K T G
polygaloides, *F. v. Mueller* ... ... A   L K T G

ELATINEAE.
Elatine, Linne (1737).
Americana, *Arnott* ... ... S   M   G

Bergia, Linne (1771).
ammannioides, *Roxburgh* ... C   M
perennis, *F. v. Mueller* ... F

LIST OF SPECIES.

HYPERICINEAE.
Hypericum, Linne.

Japonicum, *Thunberg*		F	C		A	N			K	T	G

RUTACEAE.
Correa, Smith (1798).

aemula, *F. v. Mueller*					A				K		G
alba, *Andrews*					A				K		G
speciosa, *Andrews*				W	A	N	Y	L	K	T	G
decumbens, *F. v. Mueller*					A				K		

Zieria, Smith (1798).

veronicea, *F. v. Mueller*					A				K		

Boronia, Smith (1798)

Edwardsi, *Bentham*					A				K		
coerulescens, *F. v. Mueller*				W	A	N		L	K	T	G
filifolia, *F. v. Mueller*					A				K	T	G
clavellifolia, *F. v. Mueller*				W			Y			T	
parviflora, *Smith*					A				K		
polygalifolia, *Smith*					A						G
pinnata, *Smith*										T	G

Eriostemon, Smith (1798).

Species											
obovalis, *Cunningham*										T	
linearis, *Cunningham*				W							
difformis, *Cunningham*				W	M	A			L	K	T
lepidotus, *Sprengel*											T
stenophyllus, *F. v. Mueller*											T
sediflorus, *F. v. Mueller*				W	M		N	Y			T
pungens, *Lindley*						A			L		
Hillebrandi, *F. v. Mueller*						A					
brachyphyllus, *F. v. Mueller*						A			L		
capitatus, *F. v. Mueller*				W			N	Y	L	K	

Geijera, Schott (1834).

salicifolia, *Schott*					M						
parviflora, *Lindley*			S	W	M	·	N	Y	L	K	

MELIACEAE.
Owenia, F. v. Mueller (1857).

acidula, *F. v. Mueller*		F									

LINEAE.
Linum, Linne.

marginale, *Cunningham*				M	A	N	Y	L		T	G

ZYGOPHYLLEAE.
Tribulus, Linne (1735).

terrestris, *Linne*	...	...	F	C	S	W	M				T
hystrix, *R. Brown*	...	...	F	C							
macrocarpus, *F. v. Mueller*	...	...	F								
Forrestii, *F. v. Mueller*	...	...	F								
platypterus, *Bentham*	...	...	F								
hirsutus, *Bentham*	...	...	F								
astrocarpus, *F. v. Mueller*	...	...	F								

Zygophyllum, Linne (1735).

apiculatum, *F. v. Mueller*	...	F		S	W	M		N			
fruticulosum, *DeCandolle*	...	F	C	S	W	M	.	N	Y		
ammophilum, *F. v. Mueller*	...	F	C		W	M			Y	L	K	T
Billardieri, *De Candolle*	...		C	S	W	M	A	?	Y	L	K	T	G
prismatothecum, *F. v Mueller*	...			S								
Howittii, *F. v. Mueller*	...		C	S								
glaucescens, *F. v. Mueller*	...		C		W	M	A	N	Y	L		
crenatum, *F. v. Mueller*	...				W	M		N	Y			
iodocarpum, *F. v. Mueller*	...	F	C	S	W	M						

Nitraria, Linne (1759).

Schoeberi, *Linne* ... ... C S W M A N Y L K T

GERANIACEAE.
Pelargonium, L'Heritier (1787).

australe, *Willdenow*	...	...	S			A	N	Y	L	K	T	G
Rodneyanum, *Mitchell*	...	...	...	...	...	A					T	G

Geranium, Linne.
pilosum, *Solander* ... ... S W M A N Y L K T G

Erodium, L'Heritier (1787).
cygnorum, *Nees* ... ... F C S W M A N Y L K T G

Oxalis, Linne (1737).
corniculata, *Linne* ... F . S W M A N Y L K T G

SAPINDACEAE.
Diplopeltis, Endlicher (1837).
Stuartii, *F. v. Mueller* ... ... F

Atalaya, Blume (1847).
hemiglauca, *F. v. Mueller* ... F C

LIST OF SPECIES.   211

Heterodendron, Desfontaines (1818).

	F	C	S	W	M	A	N	Y	L	K	T	G
oleaefolium, *Desfontaines*	F	C	S	W	M		.	N	Y			

Dodonaea, Linne (1737).

	F	C	S	W	M	A	N	Y	L	K	T	G
viscosa, *Linne*	F	C	S	W	M	A	N	Y	L	K	T	G
petiolaris, *F. v. Mueller*	F											
lanceolata, *F. v. Mueller*	F											
procumbens, *F. v. Mueller*												G
lobulata, *F. v. Mueller*			S	W	M		N					
bursarifolia, *Behr & F. v. M.*					M	A		Y		K		
Baueri, *Endlicher*					M		N	Y	L	K		G
hexandra, *F. v. Mueller*						A		Y	L			
humilis, *Endlicher*				W				Y	L	K	T	
boronifolia, *G. Don*								Y				
macrozyga, *F. v. Mueller*	F											
tenuifolia, *Lindley*			S									
stenozyga, *F. v. Mueller*				W				Y			T	
microzyga, *F. v. Mueller*	F	C	S	W			N					

STACKHOUSIEAE.

Stackhousia, Smith (1798).

	F	C	S	W	M	A	N	Y	L	K	T	G
megaloptera, *F. v. Mueller*	F											
spathulata, *Sieber*										K		G
linarifolia, *Cunningham*					M	A	N	Y	L	K	T	G
flava, *Hooker*						A		Y	L	K		
muricata, *Lindley*	F											
viminea, *Smith*	F											

Macgregoria, F. v. Mueller (1873).

	F	C	S	W	M	A	N	Y	L	K	T	G
racemigera, *F. v. Mueller*	F											

PHYTOLACCEAE.

Didymotheca, J. Hooker (1847).

	F	C	S	W	M	A	N	Y	L	K	T	G
thesioides, *Hooker*								Y	L	K	.	G
pleiococca, *F. v. Mueller*	F				M	A				K	T	G

Gyrostemon, Desfontaines (1820).

	F	C	S	W	M	A	N	Y	L	K	T	G
ramulosus, *Desfontaines*	F			W								

Codonocarpus, Cunningham (1830).

	F	C	S	W	M	A	N	Y	L	K	T	G
pyramidalis, *F. v. Mueller*			S									
cotinifolius, *F. v. Mueller*	F	C			M		N					

FLORA OF SOUTH AUSTRALIA.

MALVACEAE.

Plagianthus, R. & G. Forster (1776).

Berthae, *F. v. Mueller*... ... ... Y
spicatus *Bentham* ... ... M Y K T G
glomeratus, *Bentham* ... ... F C S W Y L
microphyllus, *F. v. Mueller* ... ...A N . L T G

Sida, Linne (1737).

corugata, *Lindley* ... ... F C S W M A N Y
intricata, *F. v. Mueller* ... C N
virgata, *Hooker*... ... ... F C S W
cardiophylla, *F. v. Mueller* ... F
cryphiopetala, *F. v. Mueller* ... F
petrophila, *F. v. Mueller* ... F C S W N
calyxhymenia, *J. Gay*... ... ... W
rhombifolia, *Linne* ... ... F
inclusa, *Bentham* ... ... F C
platycalyx, *F. v. Mueller* ... F
lepida, *F. v Mueller* ... ... F

Howittia, F. v. Mueller (1855).

trilocularis, *F. v. Mueller* ... ... T

Abutilon, Gaertner (1791).

tubulosum, *Hooker* ... ... F
leucopetalum, *F. v. Mueller* ... C
Mitchelli, *Bentham* ... ... F S
cryptopetalum, *F. v. Mueller*... F
otocarpum, *F v. Mueller* ... F C M
Avicennae, *Gaertner* ... ... C M
oxycarpum, *F. v. Mueller* ... C
Fraseri, *Hooker*... ... ... F C S
halophilum, *F. v. Mueller* ... C S
macrum, *F. v. Mueller*... ... F W

Lavatera, Linne.

plebeia, *Sims* ... ... F C S W M A N Y L K T G

Malvastrum, Asa Gray (1849).

spicatum, *A. Gray* ... ... F C S W

Hibiscus, Linne (1735).

trionum, Linne C
brachysiphonius, *F. v. Mueller* C
microchlaenus, *F v. Mueller* ... F . . W
Pinonianus, *Gaudichd.*... ... F C
Krichauffii, *F. v. Mueller* ... C S W M
Sturtii, *Hooker*... ... ... F C

LIST OF SPECIES.

Farragei, *F. v. Mueller* ... ... F W
Wrayae, *Lindley* ... ... W   N Y L
hakeaefolius, *Giord.* ... ... S W   L

Gossypium, Linne (1737).
australe, *F. v. Mueller* ... ... F
Sturtii, *F. v. Mueller* ... ... F C S

TILIACEAE.
Triumfetta, Linne.
Winneckeana, *F. v. Mueller* ... F

Corchorus, Linne.
sidoides, *F. v. Mueller* ... ... F
Elderi, *F. v. Mueller* ... ... F

STERCULIACEAE.
Hermannia, Linne.
Gilesii, *F. v. Mueller* ... ... F

Waltheria, Linne (1737).
Indica, *Linne* ... ... F

Melhania, Forskael (1775).
incana, *Heyne* ... ... F

Commergonia, R. & G. Foster (1776).
magniflora, *F. v. Mueller* ... F
loxophylla, *F. v. Mueller* ... F
Kempeana, *F. v. Mueller* ... F
Tatei, *F. v. Mueller* L

Brachychiton, Schott & Endlicher (1832).
Gregorii, *F. v. Mueller* ... ... F

Seringia, Sprengel (1818).
corollata, *Steetz* ... ... F
nephrosperma, *F. v. Mueller* ... F
integrifolia, *F. v. Mueller* ... F

Hannafordia, F. v. Mueller (1860).
Bissillii, *F. v. Mueller* ... ... F

Thomasia, J. Gay (1821).
petalocalyx, *F. v. Mueller* ... ... A . Y L K T

Lasiopetalum, Smith (1798).

discolor, *Hooker*...	...	...	...	W	A		Y	L	K	G
Behrii, *F. v. Mueller* ...	...	...	...		M A	N	Y	L	K
Baueri, *Steetz* ...	...	...	...	W .	A .		Y	L	K T
Tepperi, *F. v. Mueller*...	...	...	...				Y		
Schulzenii, *F. v. Mueller*	...	...	...	...	...			L	K T G

TREMANDREAE.
Tetratheca, Smith (1793).

ciliata, *Lindley* ...	...	...	...	...	...	...	...	G
ericifolia, *Smith*...	...	...	...	...	...A		K	G

EUPHORBIACEAE.
Euphorbia, Linne (1737).

erythrantha, *F. v. Mueller*	...		C	S	W	M			
Drummondii, *Boissier*	...	F	C	S	W	M	A	N Y	T
Wheeleri, *Baillon*	...		C						
eremophila, *Cunningham*	...	F	C	S	W	M	A	N	

Poranthera, Rudge (1811).

microphylla, *Brongniart*	...	...	...	M A		Y	L K T G
ericoides, *Klotzsch*	...	...	...	...A			L K

Micrantheum, Desfontaines (1818).

hexandrum, *Hooker*	...	...	...	...	...A	K

Pseudanthus, Sieber (1837).

micranthus, *Bentham* ...	...	...	...	...A	

Phyllanthus, Linne (1737).

thesioides, *Bentham*	...	...	F				
rigens, *J. Mueller*	...	...	...	S			
rhytidospermus, *F. v. Mueller*	F	.	S				
Tatei, *F. v. Mueller*	...	...	...	...	...A	N	
calycinus, *Labillardiere*	...	...	S		N	L	T
Fuernrohrii, *F. v. Mueller*	...	C	S		M		
trachyspermus, *F. v. Mueller*...	...	...		M			
lacunarius, *F. v. Mueller*	...	F	C	S	M		
australis, *J. Hooker*	...	...	...	...	...	...	K
thymoides, *Sieber*	...	...	...	...A		K	T
Gunnii, *J. Hooker*	...	...	...	S	A N		

Amperea, A. de Jussieu (1824).

spartioides, *Brongniart*	...	...	...	...	...	...	G

Monotaxis, Brongniart (1829).

luteiflora, *F. v. Mueller*	... F

LIST OF SPECIES.

Beyeria, Miquel (1844).

viscosa, *Miquel*		A	
opaca, *F. v. Mueller*	W M	A . Y L K T G	
uncinata, *F. v Mueller*	M		

Ricinocarpus, Desfontaines (1817).

pinifolius, *Desfontaines*...

Bertya, Planchon (1845).

Mitchelli, *J. Mueller*	M A	
rotundifolia, *F. v. Mueller*		K

Adriana, Gaudichaud (1825).

quadripartita, *Gaudichaud*	M A . Y L K T G
tomentosa, *Gaudichaud*	F	M

PORTULACEAE.

Portulaca, Linne.

oleracea, *Linne*	F C S	M		
australis, *Endlicher*	F			
filifolia, *F. v. Mueller*	F C			
bicolor, *F. v. Mueller*	F			

Claytonia, Linne (1737).

pleiopetala, *F. v. Mueller*	C W		
Balonnensis, *F. v. Mueller*	F C S W		
polyandra, *F. v. Mueller*	S W		
volubilis, *F. v. Mueller*	S W M A N Y L K T		
ptychosperma, *F. v. Mueller*	C S		
brevipedata, *F. v. Mueller*	A	L	
calyptrata, *F. v. Mueller*	A	Y L K	
pumila, *F. v. Mueller*	F		
corrigiolacea, *F. v. Mueller*	F	M	Y L
Australasica, *Hooker*	A		G
pygmaea, *F. v. Mueller*	M A	Y	T

CARYOPHYLLEAE.

Saponaria, Linne (1737).

tubulosa, *F. v. Mueller*...   S   M A N Y   T

Stellaria, Linne (1753).

pungens, *Brongniart*			G
glauca, *Withering*	M A	K	G
multiflora, *Hooker*	M . N		G

Drymaria, Willdenow (1819).

filiformis, *Bentham*   M   Y

Sagina, Linne (1737).

apetala, *Linne*		S	M A N Y	. K	. G

Colobanthus, Bartling (1830).

Billardieri, *Fenzl*	G

Spergularia, Persoon (1805)·

rubra, *Cambessedes*		S W	M A N Y L K T
marina, *Wahlenb.*			M A N Y . K T

Polycarpon, Linne (1758).

tetraphyllum, *Linne*	A	L

Polycarpaea, Lamarck (1792).

synandra, *F. v. Mueller*		C
Indica, *Lamarck*	F	C

ILLECEBRACEAE.
Herniaria, Linne.

incana, *Lamarck*	S	. M

Scleranthus, *Linne* (1737).

pungens, *R. Brown*		W M A N	. L	. T
diander, *R. Brown*					G

POLYGONACEAE.
Rumex, Linne (1737).

Brownii, *Campdera*			A	Y L K	G
flexuosus, *Solander*			M A		
crystallinus, *Lange*	F	C	M		
bidens, *R. Brown*			M A		G

Polygonum, Linne.

plebeium, *R. Brown*	F	C	M A	T	
prostratum, *R. Brown*			M . N		G
lapathifolium, *Linne*					G
hydropiper, *Linne*					G
minus, *Hudson*	S		M A .		G
attenuatum, *R. Brown*		C			

Muehlenbeckia, Meissner (1840).

adpressa, *Meissner*		W	. A	Y L K	. G
Cunninghamii, *F. v. Mueller*	F C S	M A		L	T
polygonoides, *F. v. Mueller*			M		

LIST OF SPECIES.

CHENOPODIACEAE.

Atriplex, Linne.

stipitatum, *Bentham*				S	W	M						
paludosum, *R. Brown*					W	.	A	N	Y	L	K	
nummularium, *Lindley*		F	C	S	W	M						
cinereum, *Poiret*						M	A	N	Y	L	K	T	G
vesicarium, *Heward*		F	C	S	W	M	.	N				
rhagodioides, *F. v. Mueller*			C	S	.	M						
incrassatum, *F. v. Mueller*			C									
velutinellum, *F. v. Mueller*		F	C	S	W	M						
fissivalve, *F. v. Mueller*			C	S	W							
Quinii, *F. v. Mueller*			C	S								
angulatum, *Bentham*			C	S	.	M						
semibaccatum, *R. Brown*						M	A					
Muelleri, *Bentham*		F	.	S		M			Y			
prostratum, *R. Brown*				S		M	.	N	Y	L	K	
leptocarpum, *F. v. Mueller*		F	C	S		M						
limbatum, *Bentham*		F		S		M						
crystallinum, *Hooker*												G
halimoides, *Lindley*			C	S	W	M						
holocarpum, *F. v. Mueller*		F	C	S	W	M						

Dysphania, R. Brown (1810).

plantaginella, *F. v. Mueller*	F	.	.	W	
simulans, *F. v. Mueller & Tate*	F	C			
litoralis, *R. Brown*		C	S	.	M

Rhagodia, R. Brown (1810).

Billardieri, *R. Brown*					M	A	.	Y	L	T	G
parabolica, *R. Brown*			S		M	A	N	Y		T	
Gaudichaudiana, *Moquin*				W	M						
crassifolia, *R. Brown*		C		W	M			Y	L	K	T
Preissii, *Moquin*				W	.			Y			
spinescens, *R. Brown*	F	C	S	W	M			Y			
nutans, *R. Brown*	F	C	S	.	M	A		Y	L	K	T

Chenopodium, Linne.

nitrariaceum, *F. v. Mueller*			S	W	M	A	.	Y			
auricomum, *Lindley*	F	C	S	.	M						
microphyllum, *F. v. Mueller*					M	A	N	Y	L		
rhadinostachyum, *F. v. Mueller*	F										
carinatum, *R. Brown*	F	C	S	W	M	A	N	Y	.	K	.	G
cristatum, *F. v. Mueller*		C	S	W	M	.	N				
atriplicinum, *F. v. Mueller*			S		M						

Enchylaena, R. Brown (1810).

tomentosa, *R. Brown*	F	C	S	W	M	A	N	Y	L	K	.	G

Threlkeldia, R. Brown (1810).

diffusa, *R. Brown* ... ... A   K

Kochia, Roth (1799).

	F	C	S	W	M	A	N	Y	L	K	T
fimbriolata, *F. v. Mueller*			S	W							
lanosa, *Lindley*	F	C	S	W	M						
lobiflora, *F. v. Mueller*			S	W							
oppositifolia, *R. Brown*				W	.	A	N	Y	L	K	T
brevifolia, *R. Brown*	F	.	S	W	M	A	N	Y	L		
triptera, *Bentham*	F	C	S		M						
decaptera, *F. v. Mueller*				W							
pentatropis, *Tate*			S								
pyramidata, *Bentham*	F		S		M						
eriantha, *F. v. Mueller*	F	C	S	W	.		N				
spongiocarpa, *F. v. Mueller*	F										
villosa, *Lindley*	F	C	S	W	M	A		Y			T
sedifolia, *F. v. Mueller*		C	S	W	M						
aphylla, *R. Brown*	F	C	S	W	M	A					
humillima, *F. v. Mueller*					M			Y			
ciliata, *F. v. Mueller*		C	S	W	M						
brachyptera, *F. v. Mueller*		C	S	.	M	.	N				
stelligera, *F. v. Mueller*					M						

Bassia, Allioni (1766).

	F	C	S	W	M	A	N	Y	L
salsuginosa. *F. v. Mueller*					M		N		
enchylaenoides, *F. v. Mueller*						A	N	Y	
Dallachyana, *Bentham*			S		M				
tricornis, *Bentham*			S	.	M				
biflora, *F. v. Mueller*			S	.	M	A	N		L
paradoxa, *F. v. Mueller*	F	.	S	W	M		N		
lanicuspis, *F. v. Mueller*	F	C	S	W					
diacantha, *F. v. Mueller*	F	C	S	W	M	A	N	Y	L
uniflora, *F. v. Mueller*		C	S	W					
bicornis, *F. v. Mueller*	F	C	S						
eriochiton, *Tate*	F	.	S	W	M				
Cornishiana, *F. v. Mueller*		C							
quinquecuspis, *F. v. Mueller*	F	C	S	.	M				
echinopsila, *F. v. Mueller*			S	.	M				
divaricata, *F. v. Mueller*		C	S	W	M				
bicuspis, *F. v. Mueller*		C	S						

Babbagia, F. v. Mueller (1858).

	F	C	S
dipterocarpa, *F. v. Mueller*	F	C	S
acroptera, *F. v. Mueller & Tate*		C	S
pentaptera, *F. v. Mueller & Tate*		.	S

LIST OF SPECIES. 219

Salicornia, Linne.

robusta, *F. v. Mueller* ... ... M
arbuscula, *R. Brown* ... ... C S W M A N Y L K T G
australis, *Solander* ... ... C S W . A . Y L K T G
tenuis, *Bentham*... ... ... S
leiostachya, *Bentham* ... ... F C

Salsola, Linne (1737).

Kali, *Linne* ... ... F C S W M A N Y L . T G

Suaeda, Forskael (1779).

maritima, *Dumortier* ... ... M A . Y L K . G

AMARANTACEAE.
Euxolus, Rafinesque (1836).

Mitchelli, *F. v. Mueller* ... F C S M

Polycnemon, Linne (1742).

pentandrum, *F. v. Mueller* ... ... A K T G
diandrum, *F. v. Mueller* ... C W L
mesembrianthemum, *F. v. M.* C

Ptilotus, R. Brown (1810).

obovatus, *F. v. Mueller* ... F C S W M . N
incanus, *Poiret* ... ... F C S
exaltatus, *Nees* ... ... C S W
Beckeri, *F. v. Mueller* ... ... L K
gomphrenoides, *Moquin* L
helipteroides, *F. v. Mueller* ... F .
erubescens, *Schlechtendal* ... ... M A N T
alopecuroideus, *F. v. Mueller* ... F C S W M A
nobilis, *F. v. Mueller* ... F . S M A N
macrocephalus, *Poiret* ... ... T G
spathulatus, *Poiret* ... ... W M A N Y L . T G
hemisteirus, *F. v. Mueller* ... F W
Schwartzii, *F. v. Mueller* ... F
leucocoma, *Moquin* ... ... F C
parvifolius, *F. v. Mueller* ... C
Hoodii, *F. v. Mueller* ... ... F
Murrayi, *F. v. Mueller*... ... C
latifolius, *R. Brown* ... ... F C

Achyranthes, Linne (1737).

aspera, *Linne* ... ... F

Alternanthera, Forskael (1775).

triandra, *Lamarck* ... ... F C S M A K G
nana, *R. Brown*... ... ... F

FLORA OF SOUTH AUSTRALIA.

Gomphrena, Linne (1737).

Brownii, *Moquin* ... ... F

PLUMBAGINEAE.
Plumbago, Linne.

Zeylanica, *Linne* ... ... F

NYCTAGINEAE.
Boerhaavia, Linne.

diffusa, *Linne* ... ... F C S W M A N
repanda, *Willdenow* ... ... C S

URTICACEAE.
Trema, Loureiro.

cannabina, *Loureiro* ... ... F

Ficus, Linne.

platypoda, *Cunningham* ... F
orbicularis, *Cunningham* ... F

Parietaria, Linne.

debilis, *G. Forster* ... ... F C S W M A N Y L K T G

Urtica, Linne.

incisa, *Poiret* ... ... M A   K   G

CASUARINEAE.
Casuarina, Linne (1737).

quadrivalvis, *Labillardiere* ... ... S W   A   Y L K T G
glauca, *Sieber* ... ... S   M
lepidophloia, *F. v. Mueller* ... ... M
suberosa, *Otto & Dietrich* ... ... T
bicuspidata, *Bentham* L
Decaisneana, *F. v. Mueller* ... F
humilis, *Otto & Dietrich* ... ... W
distyla, *Ventenat* ... ... S   A N Y L K T G

LEGUMINOSAE.
Brachysema, R. Brown (1811).

Chambersii, *F. v. Mueller* ... F

LIST OF SPECIES.

Isotropis, Bentham (1837).

atropurpurea, *F. v. Mueller* ...	F	
Wheeleri, *F. v. Mueller* ...	F C	
Winneckei, *F. v. Mueller* ...	F	

Gompholobium, Smith (1798).

minus, *Smith* ... ... A . . L K T G

Burtonia, R. Brown (1811).

polyzyga, *Bentham* ... ... F

Mirbelia, Smith (1805).

oxyclada, *F. v. Mueller* ... F

Sphaerolobium, Smith (1805).

vimineum, *Smith* ... ... A   T G

Viminaria, Smith (1804).

denudata, *Smith* ... ... A N   T G

Daviesia, Smith (1798).

arthropoda, *F. v. Mueller* ...	F
corymbosa, *Smith* ... ... A	K
horrida, *Meissner* ... ... N	
pectinata, *Lindley*	L
ulcina, *Smith* ... ... S A N	T G
genistifolia, *Cunningham* ... ... N Y L K T	
incrassata, *Smith* ... ... A	L K
brevifolia, *Lindley* ... ... A	L K T G

Aotus, Smith (1805).

villosa, *Smith* ... ... G

Phyllota, DeCandolle (1825).

pleurandroides, *F. v. Mueller*... ... ... A . K   G
Sturtii, *Bentham* ... ... F

Eutaxia, R. Brown (1811).

empetrifolia, *Schlechtendal* ... ... M A N Y L K T G

Dillwynia, Smith (1805).

hispida, *Lindley* ... ... A	L		G
ericifolia, *Smith*... ... ... A		T	G
floribunda, *Smith* ... ... A	K		G
cinerascens, *R. Brown*			G
patula, *F. v. Mueller*	L	T	

Gastrolobium, R. Brown (1811).

elachistum, *F. v. Mueller* ... ...   W
grandiflorum, *F. v. Mueller* ... F

Pultenaea, Smith (1793).

daphnoides, *Wendland*...	...	...	...	...A		K		
stricta, *Sims*	...	...	...	...	...	...	...	G
mucronata, *F. v. Mueller*	...	...	...	...A				
scabra, *R. Brown*	...	...	...	...	...	...K		
mollis, *Lindley* ...	...	...	...	...	...	L		
rigida, *R. Brown*	...	...	...	...	...	L K		
acerosa, *R. Brown*	...	...	...	...A		L K		
vestita, *R. Brown*	...	...	...	...	...	L	T	
canaliculata, *F. v. Mueller*	...	...	...	...A		L K		
largiflorens, *F. v. Mueller*	...	...	...	...A				
laxiflora, *Bentham*	...	...	...	...A				
prostrata, *Bentham*	...	...	...	...	...	...	T	
involucrata, *Bentham* ...	...	...	...	...A		K		G
pedunculata, *Hooker*	...	...	...	...A		L		
humilis, *Bentham*	...	...	...	...	...	...	...	G
graveolens, *Tate*	...	...	...	...A				
tenuifolia, *R. Brown* ...	...	...	...	...?	Y ?	K		G
densifolia, *F. v. Mueller*	...	...	...	...A		L	T	
villifera, *Sieber* ...	...	...	...	...A		L		
viscidula, *Tate* ...	...	...	...	...	...	...K		

Platylobium, Smith (1794).

obtusangulum, *Hooker*...	...	...	...	...A		K	G
triangulare, *R. Brown*...	...	...	...	...	...	...	G

Bossiaea, Ventenat (1800).

prostrata, *R. Brown*	...	...	...	...A			G
cinerea, *R. Brown*	...	...	...	...	...	...	G
riparia, *Cunningham*	...	...	...	...	...	L	
Battii, *Tate*	...	...	..	...	W		
Walkeri, *F. v. Mueller*...	...	...	...	W			

Templetonia, R. Brown (1812).

retusa, *R. Brown*	...	...	...	S W	N	L K	
Muelleri, *Bentham*	...	...	...	...	...	...	G
aculeata, *Bentham*	...	...	...	S			
egena, *Bentham* ...	...	...	F	S W M			
sulcata, *Bentham*	...	...	...	... M	Y		

Hovea, R. Brown (1812).

longifolia, *R. Brown*	...	...	...	...	N	T
heterophylla, *Cunningham*	...	...	..	...	...	T G

Nematophyllum, F. v. Mueller (1857).

Hookeri, *F. v. Mueller*... ... F

Goodia, Salisbury (1806).

lotifolia, *Salisbury*	...	...	...	...	...A				G
medicaginea, *F. v. Mueller*	...	...	W	A	N	Y	L	K .	G

LIST OF SPECIES.

Ptychosema, Bentham (1839).

anomalum, *F. v. Mueller*	...	F
trifoliolatum, *F. v. Mueller*	...	C

Crotalaria, Linne.

linifolia, *Linne*	...	...	F
Mitchelli, *Bentham*	...	...	F C
Cunninghamii, *R. Brown*	...	...	F C
dissitiflora, *Bentham*	...	...	F C S
medicaginea, *Lamarck*	...	...	F
incana, *Linne*	...	...	F

Æschynomene, Linne (1737).

Indica, *Linne*	...	...	F C

Glycyrrhiza, Linne.

psoralcoides, *Bentham*	...	...	...	M

Indigofera, Linne.

linifolia, *Retzius*	...	...	F		
monophylla, *DeCandolle*	...	...	F		
enneaphylla, *Linne*	...	...	F		
viscosa, *Lamarck*	...	...	F C		
hirsuta, *Linne*	...	...	F C		
australis, *Willdenow*	...	...	F	A N	G
brevidens, *Bentham*	...	...	F C		
coronillifolia, *Cunningham*	...	...	F		

Tephrosia, Persoon (1807).

purpurea, *Persoon*	...	...	F
sphaerospora, *F. v. Mueller*	...	...	F

Sesbania, Persoon (1807).

aculeata, *Persoon*	...	...	F C M

Clianthus, Banks & Solander (1832).

Dampieri, *Cunningham*	...	...	C S W

Swainsonia, Salisbury (1806).

Greyana, *Lindley*	...	...	...	.. M		
coronillifolia, *Salisbury*	...	...	C			
colutoides, *F. v. Mueller*	...	...	...	W		
phacoides, *Bentham*	...	...	F C S	M A . Y		
Burkittii, *F. v. Mueller*	...	...	...	W		
oligophylla, *F. v. Mueller*	...	...	F C			
Burkei, *F. v. Mueller*	...	...	F			
oroboides, *F. v. Mueller*	...	...	. C			
campylantha, *F. v. Mueller*	...	...	C S			

224 FLORA OF SOUTH AUSTRALIA.

procumbens, *F. v. Mueller*	...	C	.		M	A	N	.			G
stipularis, *F. v. Mueller*	...	C	S	W	M	A	N				
Oliverii, *F. v. Mueller* ...	...	...		W							
lessertiifolia, *DeCandolle*	...	...		W	M	A	N	Y	K	T	G
unifoliolata, *F. v. Mueller*	...	F									
microphylla, *A. Gray* ...	...	F		W	M					T	
laxa, *R. Brown* ...	...	F			M			Y			

Lespedeza, A. Richard (1803).

lanata, *Bentham* ... ... F

Psoralea, Linne (1742).

adscendens, *F. v. Mueller*	...	...	...	...	...A			G
parva, *F. v. Mueller*	...	...	...	...	...A			
patens, *Lindley* ...	...	F	C	S	W	M	A	N	G
eriantha, *Bentham*	...	F		S	.	M			
balsamica, *F. v. Mueller*	...	F							
leucantha, *F. v. Mueller*	...	F							

Trigonella, Linne (1737).

suavissima, *Lindley* ... ... F C S

Lotus, Linne (1737).

corniculatus, *Linne*	...	...	...	...	...	...A					G
australis, *Andrews*	...	...	F	C	S	W	M	A	N	Y	L K T G

Kennedya, Ventenat (1804).

monophylla, *Ventenat* ...	...	...	...	...A		Y	L	K
prostrata, *R. Brown*	...	...	...	...A		Y	L	K T G
prorepens, *F. v. Mueller*	...	F						

Glycine, Linne (1737).

clandestina, *Wendland*...	...	F	S	W	M	A	N	Y ?	G
Latrobeana, *Bentham* ...	...	...	...	...	...A				G
falcata, *Bentham*	...	...	C						
tabacina, *Bentham*	...	...	C	S	W		N		
sericea, *Bentham*	...	...	F	C		M			
tomentosa, *Bentham*	...	...	C						

Erythrina, Linne (1737).

vespertilio, *Bentham* ... ... F

Rhynchosia, Loureiro (1790).

minima, *DeCandolle* ... ... F . S

Galactia, P. Browne (1756).

tenuiflora, *Wight & Arnott* ... F

Vigna, Savi (1824).

lanceolata, Bentham ... ... F C

LIST OF SPECIES.

Cassia, Linne.

Species	F	C	S	W	M	A	N	Y	L	K	T
Sophera, *Linne*	F	C	S								
venusta, *F. v. Mueller*	F										
notabilis, *F. v. Mueller*	F										
pleurocarpa, *F. v. Mueller*	F	C									
glutinosa, *DeCandolle*	F										
pruinosa, *F. v. Mueller*		C									
desolata, *F. v. Mueller*	F	C	S								
Sturtii, *R. Brown*		C	S	W	M			Y			
artemisioides, *Gaud.*	F		S	W							
eremophila, *Cunningham*	F	C	S	W	M	A	N	Y	L		.
circinata, *Bentham*		...	...		M						
phyllodinea, *R. Brown*	F	C	S	W	M	A	N				T

Petalostylis, R. Brown (1849).

Species	F	C	S
labicheoides, *R. Brown*	F	C	S

Bauhinia, Linne.

Species	F	C
Leichhardtii, *F. v. Mueller*	F	
Carronii, *F. v. Mueller*	.	C

Neptunia, Loureiro (1790).

monosperma, *F. v. Mueller*	F
gracilis, *F. v. Mueller*	F

Acacia, Willdenow.

Species	F	C	S	W	M	A	N	Y	L	K	T
continua, *Bentham*				W		.	A		N		
Peuce, *F. v. Mueller*		C									
spinescens, *Bentham*					M	A	.	Y	L	K	
colletioides, *Cunningham*				W	M						
genistioides, *Cunningham*				W							
rupicola, *F. v. Mueller*						A	N	.	L	K	
tetragonophylla, *F. v. Mueller*	F	C	S	W	M						
spondylophylla, *F. v. Mueller*	F										
lycopodifolia, *Cunningham*	F										
minutifolia, *F. v. Mueller*	F										
calamifolia, *Sweet*			S	W	M	A	N			K	
scirpifolia, *Meissner*	F			W							
juncifolia, *Bentham*	F										
rigens, *Cunningham*						A	.	Y	L	.	T
gonophylla, *Bentham*				W							
sessiliceps, *F. v. Mueller*	F										
papyrocarpa, *Bentham*				W							
Gilesiana, *F. v. Mueller*				W							
armata, *R. Brown*					M	A	N		L	K	T
strongylophylla, *F. v. Mueller*	F										
Sentis, *F. v. Mueller*	F	C	S								
aspera, *Lindley*					M						
acanthoclada, *F. v. Mueller*				W							

FLORA OF SOUTH AUSTRALIA.

vomeriformis, *Cunningham*					A					T	
erinacea, *Bentham*				W							
obliqua, *Cunningham*					M	A	N				
lineata, *Cunningham*									L		
sublanata, *Bentham*			S	W			N				
pravifolia, *F. v. Mueller*			S				N				
acinacea, *Lindley*					A		N	Y	L		
anceps, *DeCandolle*									L		
dodonaeifolia, *Willdenow*									L	K	T .
microcarpa, *F. v. Mueller*					M	.		Y	L	K	
brachybotrya, *Bentham*					M	A	N	Y	.	K	T
Spilleriana, *J. E. Brown*					A						
suaveolens, *Willdenow*											G
iteaphylla, *F. v. Mueller*			S								
Murrayana, *F. v. Mueller*	C										
notabilis, *F. v. Mueller*				W	M	A	N	Y	L	K	
retinodes, *Schlechtendal*					A		N		L	K	G
Wattsiana, *F. v. Mueller*							N				
pycnantha, *Bentham*			S		A		N		L	K	G
hakeoides, *Cunningham*			S		M			Y			
salicina, *Lindley*	F .			W	M	A	N	Y	L	K	
pyrifolia, *DeCandolle*	F										
myrtifolia, *Willdenow*					A				L	K	T G
verniciflua, *Cunningham*					A						
montana, *Bentham*					M	A					T
impressa, *F. v. Mueller*	F										
estrophiolata, *F. v. Mueller*	F										
craspedocarpa, *F. v. Mueller*	C										
cochlearis, *Wendland*				W					L		
dictyophleba, *F. v. Mueller*	F										
retivenea, *F. v. Mueller*	F										
trineura, *F. v. Mueller*					M						
cyclopis, *Cunningham*				W							
melanoxylon, *R. Brown*					A		N			T	G
homalophylla, *Cunningham*			S	W	M						
stenophylla, *Cunningham*	C				M						
Osswaldi, *F. v. Mueller*	C			W	M			Y			
coriacea, *DeCandolle*	C										
sclerophylla, *Lindley*					M			Y	L		T
farinosa, *Lindley*					M					K	
Whanii, *F. v. Mueller*					M					K	
lanigera, *Cunningham*								Y			
verticillata, *Willdenow*					A		N			K	. G
oxycedrus, *Sieber*											G
rhigiophylla, *F. v. Mueller*					M						
stipuligera, *F. v Mueller*	F										
lysiphloia, *F v Mueller*	F										
longifolia, *Willdenow*					A	.		Y	L	K	T G
Kempeana, *F. v. Mueller*	F										
acradenia, *F. v. Mueller*	F										

LIST OF SPECIES.

doratoxylon, *Cunningham*	F	C									
aneura, *F. v. Mueller*	F	C	S	W							
cibaria, *F. v. Mueller*	F										
cyperophylla, *F. v. Mueller*	F	C									
Burkitti, *F. v. Mueller*				W							
Farnesiana, *Willdenow*	F	C									
Mitchelli, *Bentham*											G
mollissima, *Willdenow*									T	G
dealbata, *Link*									T	G

THYMELEAE.
Pimelea, Banks & Solander (1788).

trichostachya, *Lindley*	F	C								
curviflora, *R. Brown*				W	M	A	N	Y	K	
simplex, *F. v. Mueller*	F	C	S	W						
phylicoides, *Meissner*				W		A		Y	K	G
octophylla, *R. Brown*					M	A	N	Y L	K	G
petraea, *Meissner*			S	W						
glauca, *R. Brown*					M	A	N	Y L	K T	G
ligustrina, *Labillardiere*						A			K	G
stricta, *Meissner*						A			K	
spathulata, *Labillardiere*						A	N		K	
humilis, *R. Brown*						A				T G
microcephala, *R. Brown*	F	C	S	W	M				L K	
serpyllifolia, *R. Brown*					M	A		Y	L K	G
clachantha, *F. v. Mueller*										T G
flava, *R. Brown*					M	A			L K	
petrophila, *F. v. Mueller*			S				N			
ammocharis, *F. v. Mueller*	F									

PROTEACEAE.
Petrophila, R. Brown (1809).

multisecta, *F. v. Mueller* ... K

Isopogon, R. Brown (1809).

ceratophyllus, *R. Brown* ... A K T G

Adenanthos, Labillardiere (1804).

sericea, *Labillardiere* ... K
terminalis, *R. Brown* ... A L K G

Conospermum, Smith (1798).

patens, *Schlechtendal* ... A L K T G
Mitchellii, *Meissner* ... G

FLORA OF SOUTH AUSTRALIA.

Persoonia, Smith (1798).

juniperina, *Labillardiere*						A				T

Grevillea, R. Brown (1809).

Huegelii, *Meissner*				W	M	A	N	Y	L	G
Treueriana, *F. v. Mueller*	F		W							
ilicifolia, *R. Brown*					M	A	N	Y	L K T	
aquifolium, *Lindley*										T
angulata, *R. Brown*	F									
Wickhami, *Meissner*	F									
agrifolia, *Cunningham*	F									
pterosperma, *F. v. Mueller*	F	C	W	M						
stenobotrya, *F. v. Mueller*	F									
juncifolia, *Hooker*	F	C	W							
halmaturina, *Tate*									L K	G
nematophylla, *F. v. Mueller*		C	S	W						
striata, *R. Brown*	F	C								
lavandulacea, *Schlechtendal*						A	N		L K T	G
aspera, *R. Brown*			W						L K	
pauciflora *R. Brown*									L K	

Hakea, Schrader (1797).

chordophylla, *F. v. Mueller*	F									
lorea, *R. Brown*	F									
macrocarpa, *Cunningham*	F									
multilineata, *Meissner*	F		W							
Baxteri, *R. Brown*			W							
Ednieana, *Tate*		S								
purpurea, *Hooker*	F			M						
vittata, *R. Brown*						A			L	T
nodosa, *R. Brown*										G
cycloptera, *R. Brown*									L	
leucoptera, *R. Brown*	F	C	S	M		Y				
rostrata, *F. v. Mueller*						A			K	G
rugosa, *R. Brown*						A	N	Y	L K T	G
ulicina, *R. Brown*						A		Y	K T	G
nitida, *R. Brown*			W							

Banksia, Linne fils (1781).

marginata, *Cavanilles*						A	N		L K T	G
ornata, *F. v. Mueller*						A			K T	G

SAXIFRAGEAE.

Bauera, Banks (1793).

rubioides, *Andrews*		A	K

CRASSULACEAE.
Tillaea, Linne.

verticillaris, *De Candolle*		C	S	W	M	A	N	Y	L	K	T	G
purpurata, *Hooker*						A	N	Y				G
recurva, *Hooker*...					M	A				K		G
macrantha, *Hook.*					M	A		Y	L	K		G

ROSACEAE.
Geum, Linne (1737).

urbanum, *Linne* ... G

Potentilla, Linne (1737).

anserina, *Linne* ... A G

Rubus, Linne.

parvifolius, *Linne* ... A K G

Acaena, Linne (1771).

ovina, *Cunningham*	... A	N	Y	L	K	T	G
Sanguisorbae, *Vahl*	... A				K		G

Stylobasium, Desfontaines (1819).

spathulatum, *Desfontaines* ... F

FICOIDEAE.
Mesembrianthemum, Linne (1737).

aequilaterale, *Haworth*	... F	. S	W	M	A	N	Y	L	K	T	G
australe, *Solander*			W	M	A	N	Y	L	K	T	

Tetragonia, Linne (1737).

expansa, *Murray*	C	S	W	M		N				
implexicoma, *Hook.*					A	N	Y	L	K	T	G

Gunnia, F. v. Mueller (1858).

septifraga, *F. v. Mueller* ... C

Aizoon, Linne (1737).

quadrifidum, *F. v. Mueller*	... C	S	W
zygophylloides, *F. v. Mueller* ...	C	S	

Trianthema, Linne.

turgidifolia, *F. v. Mueller*	... F			
crystallina, *Vahl*	...	C	S	M
pilosa, *F. v. Mueller*	... F	C		
humillima, *F. v. Mueller*	...			Y

Zaleya, Burmann (1768).
decandra, *Burmann* ... ... C W

Mollugo, Linne (1737).
hirta, *Thunberg* ... ... F C S M
orygioides, *F. v. Mueller* ... C
Spergula, *Linne* ... ... M
Cerviana, *Seringe* ... ... F . W M

LYTHRARIEAE.
Lythrum, Linne (1737).
Salicaria, *Linne* ... ...A G
hyssopifolia, *Linne* ... ... F C A G

Rotala, Linne (1771).
diandra, *F v. Mueller* ... ... F
verticillaris, *Linne* ... ... F

Ammannia, Linne (1737).
baccifera, *Linne* C
multiflora, *Roxburgh* ... ... F M

ONAGREAE.
Epilobium, Linne (1737).
glabellum, *G. Forster* ... ... S A N L K G

Jussieua, Linne (1737).
diffusa, *Forskael* ... ... M A G

MYRTACEAE.
Darwinia, Rudge (1813).
micropetala, *Bentham* ... ... W . K
Schuermanni, *Bentham* L

Verticordia, DeCandolle (1813).
Wilhelmii, *F. v. Mueller* L

Calycothrix, Labillardiere (1806).
longiflora, *F. v. Mueller* ... F
tetragona, *Labillardiere* ... ...A N Y L K T G

Lhotzkya, Schauer (1835).
glaberrima, *F. v. Mueller* ... ... K
genetylloides, *F. v. Mueller* ... ... T G
Smeatoniana, *F. v. Mueller* ... ... K

LIST OF SPECIES.

Thryptomene, Endlicher (1838).

Species	F	S	W	M	A	N	Y	L	K	T	G
Maisonneuvii, *F. v. Mueller*	F										
flaviflora, *F. v. Mueller*	F										
Mitchelliana, *F. v. Mueller*				M						T	
auriculata, *F. v. Mueller*			W								
Elliottii, *F. v. Mueller*			W								
ericaea, *F. v. Mueller*									K		
Miqueliana, *F. v. Mueller*							Y	L			
ciliata, *F. v. Mueller*									K	T	G

Baeckea, Linne (1753).

Species	F	S	W	M	A	N	Y	L	K	T	G
diffusa, *Sieber*					A				K		
crassifolia, *Lindley*							Y	L	K	T	G
ericaea, *F. v. Mueller*										T	
polystemona, *F. v. Mueller*	F										
Behrii, *F. v. Mueller*			W					L		T	

Leptospermum, R. & G. Forster (1776).

Species	F	S	W	M	A	N	Y	L	K	T	G
laevigatum, *F. v. Mueller*			W	M			Y		K	T	
scoparium, *Forster*					A	N			K		G
lanigerum, *Smith*					A				K		G
myrsinoides, *Schlechtendal*					A				K	T	G

Kunzea, Reichenbach (1828).

Species	F	S	W	M	A	N	Y	L	K	T	G
pomifera, *F. v. Mueller*					A				K	T	G

Callistemon, R. Brown (1814).

Species	F	S	W	M	A	N	Y	L	K	T	G
coccineus, *F. v. Mueller*					A	N	Y	L	K	T	G
salignus, *DeCandolle*					A					T	
teretifolius, *F. v. Mueller*		S			A						
brachyandrus, *Lindley*				M						T	

Melaleuca, Linne (1767).

Species	F	S	W	M	A	N	Y	L	K	T	G
squamea, *Labillardiere*									K		G
Wilsonii, *F. v Mueller*								L		T	G
gibbosa, *Labillardiere*									K		G
decussata, *R. Brown*					A			L	K	T	G
squarrosa, *Smith*									K	T	G
glomerata, *F. v. Mueller*	F	C	S								
trichostachya, *Lindley*		C									
parviflora, *Lindley*	F		S	W	M	A	N	Y	L	K	T	G
cylindrica, *R. Brown*									K		
acuminata, *F. v. Mueller*				M	A		Y	L	K	T	
quadrifaria, *F. v. Mueller*			W								
uncinata, *R. Brown*			W	M	A	N	Y	L	K	T	
ericifolia, *Smith*											G
pustulata, *Hooker*					A		Y	L	K	T	G

FLORA OF SOUTH AUSTRALIA.

Eucalyptus, L'Heritier (1788).

corynocalyx, *F. v. Mueller*						N		L	K	
gamophylla, *F. v. Mueller*	F									
tessellaris, *F. v. Mueller*	F									
incrassata, *Labillardiere*			S	W	M	A	N	Y	L	K	T	G
hemiphloia, *F. v. Mueller*							N		L	K	T
gracilis, *F. v. Mueller*					M	A	N	Y		K	T
odorata, *F. v. Mueller*						A	N	Y		K	T	G
pauciflora, *Sieber*												G
amygdalina, *Labillardiere*												G
obliqua, *L'Heritier*						A				K		G
Sieberiana, *F. v. Mueller*										K		G
paniculata, *Smith*						A				K		G
largiflorens, *F. v. Mueller*					M					K	T
Behriana, *F. v. Mueller*							N		L	K	T
oleosa, *F. v. Mueller*			S				N	Y			T
terminalis, *F. v. Mueller*	F										
setosa, *Schauer*	F										
goniocalyx, *F. v. Mueller*						A	N			K	
leucoxylon, *F. v. Mueller*						A	N		L	K	T	G
uncinata, *Turczaninow*					M	A		Y		K	T
cneorifolia, *DeCandolle*									L	K	
microtheca, *F. v. Mueller*	F	C									
Stuartiana, *F. v. Mueller*												G
vimninalis, *Labillardiere*						A			L	K		G
rostrata, *Schlechtendal*	F	C	S	W	M	A	N		L	K	T	G
Gunnii, *Hooker*						A						G
cosmophylla, *F. v. Mueller*						A			L	K	
santalifolia, *F. v. Mueller*						A			L	K		G
capitellata, *Smith*						A				K	T	G
macrorrhyncha, *F. v. Mueller*						A					
Oldfieldii, *F. v. Mueller*	F										
pachyphylla, *F. v. Mueller*	F										
pyriformis, *Turczaninow*	F		W								

RHAMNACEAE.

Ventilago, Gaertner (1788).

viminalis. *Hooker*	F	

Pomaderris, Labillardiere (1804).

myrtilloides, *Fenzl*		W					
apetala, *Labillardiere*				N			G
racemosa, *Hooker*			A	N	Y	L	K	G
obcordata, *Fenzl*				N	Y	L	K	G

Cryptandra, Smith (1798).

Wayii, *F. v. Mueller & Tate*			A	N	
Hookeri, *F. v. Mueller*			A			G
phlebophylla, *F. v. Mueller*	S				

spathulata, *F. v. Mueller* ... F A L K G
coactilifolia, *F. v. Mueller* ... ... A
leucophracta, *Schlechtendal* ... ... W M Y L K
obovata, *Hooker* ... ... A K
vexillifera, *Hooker* ... ... A Y L K T G
subochreata, *F. v. Mueller* ... ... M A L T G
halmaturina, *F. v. Mueller* ... ... A K
bifida, *F. v. Mueller* L
scabrida, *Tate* ... ... K
Waterhousei, *F. v. Mueller* ... ... K
hispidula, *Reisseck* ... ... A K T
propinqua, *Cunningham* ... ... M
amara, *Smith* ... ... W A
tomentosa, *Lindley* ... ... W M A Y L

OLACINEAE.
Olax, Linne (1747).
Benthamiana, *Miquel* ... ... L K

SANTALACEAE.
Santalum, Linne (1742).
lanceolatum, *R. Brown*... ... F C S
acuminatum, *DeCandolle* ... F S W M A N Y L K
persicarium, *F. v. Mueller* ... ... S M A N Y T

Choretrum, R. Brown (1810).
glomeratum, *R. Brown*... ... ... A Y L K T
chrysanthum, *F. v. Mueller* ... ... A Y
spicatum, *F. v. Mueller* ... ... M K T

Leptomeria, R. Brown (1810).
aphylla, *R. Brown* ... ... M A N L K G

Anthobolus, R. Brown (1810).
exocarpoides, *F. v. Mueller* ... F

Exocarpos, Labillardiere (1798).
cupressiformis, *Labillardiere* ... ... M A Y L K T G
spartea, *R. Brown* ... ... F M A Y L
aphylla, *R. Brown* ... ... S W M A N Y L
stricta, *R. Brown* ... ... M L T G

HALORAGEAE.
Loudonia, Lindley (1839).
Behrii, *Schlechtendal* ... ... A Y L K T
aurea, *Lindley* ... ... S W K

FLORA OF SOUTH AUSTRALIA.

Haloragis, R. & G. Forster (1776).

Species	F	C	S	W	M	A	N	Y	L	K	T	G
Meionectes, *F. v. Mueller*						A						G
heterophylla, *Brongniart*		C	S		M	A	N	Y	L			G
digyna, *Labillardiere*				W	M	A				K		G
elata, *Cunningham*				W		A				K		
aspera, *Lindley*	F	C	S	W	M	A	N	Y				G
acutangula, *F. v. Mueller*									L			
odontocarpa, *F. v. Mueller*				W	M							
trigonocarpa, *F. v. Mueller*	F											
Gossei, *F. v. Mueller*	F											
micrantha, *R. Brown*						A						G
tetragyna, *R. Brown*						A	N	Y	L	K	T	G
teucrioides, *A. Gray*						A	N	Y	L	K		G

Myriophyllum, Linne (1767).

Species	F	C	S	W	M	A	N	Y	L	K	T	G
integrifolium, *Hooker*					M	A					T	G
amphibium, *Labillardiere*						A						G
pedunculatum, *Hooker*												G
verrucosum, *Lindley*	F				M	A			L			G
Muelleri, *Sonder*						A			L	K		
intermedium, *DeCandolle*					M	A				K		G
elatinoides, *Gaudichaud*					M	A				K	T	G

Callitriche, Linne (1748).

Species	F	C	S	W	M	A	N	Y	L	K	T	G
verna, *Linne*					M	A					T	

UMBELLIFERAE.

Actinotus, Labillardiere (1804).

Species	F	C	S	W	M	A	N	Y	L	K	T	G
Schwarzii, *F. v. Mueller*	F											

Hydrocotyle, Linne.

Species	F	C	S	W	M	A	N	Y	L	K	T	G
vulgaris, *Linne*			S		M	A	N				T	G
Asiatica, *Linne*			S		M	A				K		G
Candollei, *F. v. Mueller*			S			A	N			K	T	G
hirta, *R. Brown*						A			L	K		G
pterocarpa, *F. v. Mueller*												G
comocarpa, *F. v. Mueller*										K		
tripartita, *R. Brown*										K		G
callicarpa, *Bunge*			S		M	A	N	Y	L	K	T	G
trachycarpa, *F. v. Mueller*	F		S									
crassiuscula, *Tate*										K		
capillaris, *F. v. Mueller*				W		A		Y	L	K		G
medicaginoides, *Turcz.*								Y				
diantha, *DeCandolle*										K		

LIST OF SPECIES. 235

Didiscus, DeCandolle (1828).

pusillus, *F. v. Mueller* ... ...A				Y	L	K
cyanopetalus, *F. v. Mueller* ... F		M		Y	L	
eriocarpus, *F. v. Mueller* ... ... S W				Y	L	
pilosus, *Bentham*						G
glaucifolius, *F. v. Mueller* ... F C S						

Trachymene, Rudge (1810).

heterophylla, *F. v. Mueller* ... ...A   L K

Xanthosia, Rudge (1810).

pusilla, *Bunge* ... ...A	Y				
dissecta, *Hooker* ... ...A		L	K	T	G

Eryngium, Linne.

rostratum, *Cavan.* ... ...A			T	G
vesiculosum, *Labillardiere* ... ...A		K		G
plantagineum, *F. v. Mueller* ... C				

Crantzia, Nuttall (1818).

lineata, *Nuttall* ... ... M A   K   G

Caldasia, Lagasca (1821).

andicola, *Lagasca* ... ... G

Apium, Linne.

prostratum, *Labillardiere* ... ... M A N Y L K T G

Sium, Linne.

latifolium, *Linne* ... ...A

Daucus, Linne.

brachiatus, *Sieber* ... F C S W M A N Y L K T G

CUCURBITACEAE.

Cucumis, Linne.

Chate, *Linne* ... F   M

Momordica, Linne.

Charantia, *Linne* ... ... F

Melothria, Linne.

Muelleri, *Bentham* ... ... M	
Maderaspatana, *Cogniaux* ... F S	

LORANTHACEAE.

Loranthus, Linne (1740).

celastroides, *Sieber*	...	...	...	...	...	...	...	...	G
angustifolius, *R. Brown*	...	...	...	...	...	...	L	
Exocarpi, *Behr* ...	...	...	F	C	S	W	A	Y	
linearifolius, *Hooker*	...	...	...	S					
Murrayi, *Tate* ...	...	...	...	S					
linophyllus, *Fenzl*	...	...	F	C		W	M	A	Y
gibberulus, *Tate*	...	...	F	C					
pendulus, *Sieber*...	...	...	F	C		W	A	Y	L	G
Quandang, *Lindley*	...	...	F	C	S				
grandibracteus, *F. v. Mueller*...			C						

Viscum, Linne.

articulatum, *Burm.* C

RUBIACEAE.

Oldenlandia, Linne.

tillaeacea, *F. v. Mueller* ... F C S

Dentella, Forster (1776).

repens, *Forster* ... ... F C

Canthium, Lamarck (1783).

latifolium, *F. v. Mueller* ... F

Coprosma, Forster (1776).

hirtella, *Labillardiere* ... ... G

Opercularia, Gaertner (1788).

ovata, *J. Hooker*	...	...	...	...	...	A		G
varia, *J. Hooker*	...	...	...	...	...	A	Y	K	G
scabrida, *Schlechtendal*...	...	...	...	A	L			

Pomax, Solander (1788).

umbellata, *Solander* ... ... F S W N

Spermacoce, Linne.

marginata, *Bentham* ... ... F

Asperula, Linne.

geminifolia, *F. v. Mueller*	...	...	...	M				T
oligantha, *F. v. Mueller*	...	...	S	M	A	N	Y	K	T	G

Galium, Linne.

umbrosum, *Solander*	...	...	...	W	M	A	N	Y	L	K	T	G
australe, *DeCandolle*	...	...	...	S	M	A	N	Y		K	T	G

LIST OF SPECIES.

CAPRIFOLIACEAE.
Sambucus, Linne.

Species	F	S	C	W	M	A	N	Y	L	K	T	G
Gaudichaudiana, *DeCandolle*												G

COMPOSITAE.
Siegesbeckia, Linne (1737).

Species	F	S	C	W	M	A	N	Y	L	K	T	G
orientalis, *Linne*	F			W		A	N					G

Wedelia, Jacquin (1763).

Species	F	S	C	W	M	A	N	Y	L	K	T	G
platyglossa, *F. v. Mueller*			C		M	A					T	
verbesinoides, *F. v. Mueller*	F											

Bidens, Linne (1753).

Species	F	S	C	W	M	A	N	Y	L	K	T	G
bipinnata, *Linne*	F											

Glossogyne, Cassini (1827).

Species	F	S	C	W	M	A	N	Y	L	K	T	G
tenuifolia, *Cassini*	F	S										

Flaveria, Jussieu (1789).

Species	F	S	C	W	M	A	N	Y	L	K	T	G
Australasica, *Hooker*			C									

Aster, Linne.

Species	F	S	C	W	M	A	N	Y	L	K	T	G
Sonderi, *F. v. Mueller*						A						
pannosus, *F. v. Mueller*						A	N	Y	L		T	G
pimeloides, *Cunningham*		S		W	M		N					
myrsinoides, *Labillardiere*												G
Mitchelli, *F. v. Mueller*	F				M			Y				
tubuliflorus, *F. v. Mueller*						A				K	T	
axillaris, *F. v. Mueller*						A	N	Y	L	K	T	G
microphyllus, *Ventenat*												G
ramulosus, *Labillardiere*				W	M	A		Y	L	K		G
exiguifolius, *F. v. Mueller*				W								
lepidophyllus, *Persoon*						A		Y			T	
stellulatus, *Labillardiere*												G
asterotrichus, *F. v. Mueller*												G
magniflorus, *F. v. Mueller*				W	M							
calcareus, *F. v. Mueller*		S		W	M							
Muelleri, *Sonder*			C	W	M			Y			T	
Stuartii, *F. v. Mueller*				W								
decurrens, *Cunningham*					M			Y	L			
glutescens, *F. v. Mueller*		S			M	A		Y			T	G
teretifolius, *F. v. Mueller*					M	A		Y		K		G
glandulosus, *Labillardiere*												G
megalodontus, *F. v. Mueller*	F											
Ferresii, *F. v. Mueller*	F											
exul, *Lindley*				W	M	A	N	Y	L	K		G
Huegelii, *F. v. Mueller*				W		A		Y	L	K	T	G

Podocoma, Cassini (1817).

cuneifolia, *R. Brown*			F	C	S						

Vittadinia, Richard (1832).

australis, *Richard*	F		S	W	M	A	N	Y	L	K	T	G

Dimorphocoma, F. v. Mueller & Tate (1883).

minutula, *F. v. Mueller & Tate*			S							

Minuria, DeCandolle (1836).

leptophylla, *DeCandolle*	F	C	S	W	M	A	N	Y	L	T
Cunninghamii, *Bentham*	F	C	S		M			Y		
integerrima, *Bentham*		C	S		M					
denticulata, *Bentham*	F	C	S		M					
suaedifolia, *F. v. Mueller*		C		W			N	Y	L	T

Achnophora, F. v. Mueller (1883).

Tatei, *F. v. Mueller*									K

Calotis, R. Brown (1820).

cuneifolia, *R. Brown*					M	A	N			
hispidula, *F. v. Mueller*	F	C	S	W	M		N	Y		
cymbacantha, *F. v. Mueller*	F	C	S	W			N			
erinacea, *Steetz*		C	S		M	A		Y	L	
scabiosifolia, *Sonder & F. v. M.*			S			A				
scapigera, *Hooker*					M					
lappulacea, *Bentham*	F	C	S							
microcephala, *Bentham*	F									
plumulifera, *F. v. Mueller*		C	S							
porphyroglossa, *F. v. Mueller*	F	C								
Kempei, *F. v. Mueller*	F									

Lagenophora, Cassini (1818).

Billardieri, *Cassini*				A			L	K	G
Huegelii, *Bentham*				A	N	Y	L		G

Brachycome, Cassini (1816).

goniocarpa, *Sonder & F. v. M.*				M		N		L		
pachyptera, *Turczaninow*		C	S	W	M	A	N	Y		
collina, *Bentham*						A	N	Y	L	
Muelleri, *Sonder*						A			L	T	G
graminea, *F. v. Mueller*				W	M	A				G
basaltica, *F. v. Mueller*						A				
trachycarpa, *F. v. Mueller*			S	W		A	N			G
diversifolia, *Fischer & Meyer*						A				G
ciliaris, *Lessing*	F	C	S	W	M	A	N	Y	L	T
debilis, *Sonder*						A			L	
chrysoglossa, *F. v. Mueller*					M					

LIST OF SPECIES. 239

calocarpa, *F. v. Mueller* ... ... S M A
exilis, *Sonder* ... ... M A N L
melanocarpa, *Sonder & F. v. M.* C W M
cardiocarpa, *F. v. Mueller* ... ... G
cuneifolia, *Tate* ... ... L K
decipiens, *Hooker* ... ... G

Erodiophyllum, F. v. Mueller (1875).
Elderi, *F. v. Mueller* ... ... C W

Cymbonotus, Cassini (1825).
Lawsonianus, *Gaudichaud* ... ... A N Y K G

Solenogyne, Cassini.
emphysopus, *F. v. Mueller* ... ... A K T G

Isoetopsis, Turczaninow (1851).
graminifolia, *Turczaninow* ... ... S M A N Y L K T

Toxanthus, Turczaninow (1851).
perpusillus, *Turczaninow* ... ... M N Y
Muelleri, *Bentham* ... ... M A L K

Quinetia, Cassini (1830).
Urvillei, *Cassini* ... ... A L

Millotia, Cassini (1829).
tenuifolia, *Cassini* ... ... S W M A N Y L K T G
Greevesii, *F. v. Mueller* ... C S M
Kempei, *F. v. Mueller* ... ... F S W

Erechthites, Rafinesque (1817).
prenanthoides, *DeCandolle* ... ... G
picridioides, *Turczaninow* ... F N Y K T
arguta, *DeCandolle* ... ... S A K T G
mixta, *DeCandolle* ... ... M L
quadridentata, *DeCandolle* ... ... M A N Y L K G
hispidula, *DeCandolle* ... ... M A Y L T G

Senecio, Linne.
Gregorii, *F. v Mueller* ... F C S W M
platylepis, *DeCandolle* ... ... M
spathulatus, *Richard* ... ... G
megaglossus, *F. v. Mueller* ... ... N
magnificus, *F. v. Mueller* ... F S
lautus, *Solander* ... ... F C S W M A N Y L K G
Behrianus, *Sonder & F. v.M.* ... ... M
dryadeus, *Sieber* ... ... G
anethifolius, *Cunningham* ... ... S N
odoratus, *Hornemann* ... ... F A N L K G

240   FLORA OF SOUTH AUSTRALIA.

hypoleucus, *Bentham*					A					
Cunninghamii, *DeCandolle*	C	S	W	M	A	N	Y		K	
Georgianus, *DeCandolle*					A				K	
brachyglossus, *F. v. Mueller*	C	S	W	M	A	N	Y			

Cotula, Linne (1735).

filifolia, *Thunberg*				M	A	N		L	K	T	G
coronopifolia, *Linne*	C	S		M	A	N		L	K	T	G
australis, *Hooker*				M	A	N		L	K	T	G
reptans, *Bentham*											G

Centipeda, Loureiro (1790).

orbicularis, *Loureiro*		C		M						
Cunninghamii, *F. v. Mueller*	F	C		M	A	N	Y	L	K	T	G
thespidioides, *F. v. Mueller*	F	C	S	M						

Ceratogyne, Turczaninow (1851).

obionoides, *Turczaninow* ... ... S

Ethuliopsis, F. v. Mueller (1861).

Cunninghamii, *F. v. Mueller* ... C S M

Epaltes, Cassini (1818).

australis, *Lessing*	F	C		M				
Tatei, *F. v. Mueller*							L	T

Stuartina, Sonder (1852).

Muelleri, *Sonder* ... ... S   M A N Y L K T G

Humea, Smith (1804).

squamata, *F. v. Mueller*		M		T
cassiniacea, *F. v. Mueller*			L	

Ixodia, R. Brown (1812).

achilleoides, *R. Brown* ... ...A   L K   G

Elachanthus, F. v. Mueller (1852).

pusillus, *F. v. Mueller* ... ... S W M   Y

Rutidosis, DeCandolle (1837).

helichrysoides, *DeCandolle*	F	C	S			
Pumilo, *Bentham*				A	Y	L	F

Pluchea, Cassini (1817),

conocephala, *F. v. Mueller*		W	M	N	
tetranthera, *F. v. Mueller*	F				
Eyrea, *F. v. Mueller*	F	C	S		N

LIST OF SPECIES.

Pterigeron, DeCandolle (1836).

liatroides, *Bentham*		C	S	W				
microglossus, *Bentham*...	F							
adscendens, *Bentham*	F							
dentatifolius, *F. v. Mueller*	F							

Podosperma, Labillardiere (1806).

angustifolium, *Labillardiere*				M A N Y L K T G

Ixiolaena, Bentham (1837).

leptolepis, *Bentham*		C	S		M			
supina, *F. v. Mueller*					A		L K	
tomentosa, *Sonder & F. v. M.*	F	C	W	S	M	N	L	

Athrixia, Ker (1823).

tenella, *Bentham*			S	M	N Y L	T	

Cassinia, R. Brown (1817).

aculeata, *R. Brown*				M		Y		G
arcuata, *R. Brown*				M A		Y		
laevis, *R. Brown*		S W						
punctulata, *F. v. M. & Tate*					N Y L K T	
spectabilis, *R. Brown*...					Y	K	

Podolepis, Labillardiere (1806).

rutidochlamys, *F. v. Mueller*	C							
canescens, *Cunningham*	F	C	S	W	M A N Y L	T	
acuminata, *R. Brown*					A	Y		G
rugata, *Labillardiere*					A	Y L K T G
Lessoni, *Bentham*			S		A			
Siemssenia, *F. v. Mueller*		C	S	W	M	N		

Gnaphalium, Linne (1737).

luteo-album, *Linne*	F	C	S		M A N Y L K T G
Indicum, *Linne*	F							
Japonicum, *Thunberg*	F		S		M A N Y L K T G
indutum, *Hooker*					M A N Y L K	G

Leptorrhynchos, Lessing (1832).

tenuifolius, *F. v. Mueller*								G
squamatus, *Lessing*			S		M A N Y L	T G
pulchellus, *F. v. Mueller*		C		W	M A N Y L	T G
elongatus, *DeCandolle*					A N Y	T G
medius, *Cunningham*					M A N Y L	T G
Waitzia, *Sonder*...				W	M A N Y L	

Q

Helipterum, DeCandolle (1837).

Species	F	C	S	W	M	A	N	Y	L	K	T	G
roseum, *Bentham*				W								
anthemoides, *DeCandolle*						A	N				T	
polygalifolium, *DeCandolle*			S	W	M				L			
strictum, *Bentham*		C	S	W								
hyalospermum, *F. v. Mueller*		C	S	W	M	A	N	Y	L			
floribundum, *DeCandolle*	F	C	S	W	M		N	Y				
heteranthum, *Turczaninow*								Y				
tenellum, *Turczaninow*				W								
pygmaeum, *Bentham*			S	W	M	A	N	Y	L			
corymbiflorum, *Schlechtendal*		C	S	W	M	A	N	Y				
stipitatum, *F. v. Mueller*	F											
incanum, *DeCandolle*	F	C	S		M		N					
Cotula, *DeCandolle*				W	M							
Haigii, *F. v. Mueller*				W								
laeve, *Bentham*			S									
dimorpholepis, *Bentham*						A	N	Y		K	T	
exiguum, *F. v. Mueller*		C		W	M	A	N	Y		K	T	G
moschatum, *Bentham*	F	C	S	W	M	A	N				T	
pterochaetum, *Bentham*	F	C	S	W								
Tietkensii, *F. v. Mueller*	F			W			N					
Charsleyae, *F. v. Mueller*	F											

Waitzia, Wendland (1808).

Species	F	C	S	W	M	A	N	Y	L	K	T	G
corymbosa, *Wendland*				W	M	A	N					

Helichrysum, Vaillant (1719).

Species	F	C	S	W	M	A	N	Y	L	K	T	G
Cassinianum, *Gaudichaud*	F	C		W								
Ayersii, *F. v. Mueller*	F											
Lawrencella, *F. v. Mueller*	F	C		W								
semifertile, *F. v. Mueller*	F	C	S	W								
scorpioides, *Labillardiere*						A					T	G
rutidolepis, *DeCandolle*	F					A						
lucidum, *Henckel*	F			W	M	A	N	Y	L	K		G
podolepideum, *F. v. Mueller*		C	S									
obtusifolium, *Son. & F. v. M.*						A		Y	L	K		G
Blandowskianum, *Steetz*						A						G
adenophorum, *F. v. Mueller*										K	T	
leucopsidium, *DeCandolle*						A	N	Y	L	K		G
Baxteri, *F. v. Mueller*						A		Y	L		T	G
ambiguum, *Turczaninow*	F		S			A						
Tepperi, *F. v. Mueller*						A		Y				
apiculatum, *DeCandolle*	F	C	S	W	M	A	N	Y	L	K	T	G
semipapposum, *DeCandolle*				W	M	A	N	Y			T	G
Dockerii, *F. v. Mueller*					M							
Thomsoni, *F. v. Mueller*	F											
decurrens, *F. v. Mueller*					M	A	N		L			
retusum, *Sonder & F. v. Mueller*			S		M	A	N	Y	L	K		

LIST OF SPECIES. 243

ferrugineum, *Lessing* ...								G
cinereum, *F. v. Mueller*								G
Kempei, *F. v. Mueller* ...	F							

Polycalymma, F. v. Mueller.

Stuartii, *F. v. Mueller* ...	F	C	S	W	M			T

Hyalolepis, DeCandolle.

rhizocephala, *DeCandolle*				M	N	Y	L	K T G
Rudallii, *F. v. Mueller* ...	F	C						

Angianthus, Wendland (1809).

pleuropappus, *Bentham*							L	
brachypappus, *F. v. Mueller* ...		S						
tomentosus, *Wendland* ...	F	S		M	A	N Y	L	T
pusillus, *Bentham*	F	C	S	W	M			
tenellus, *Bentham*							L	

Skirrophorus, DeCandolle.

strictus, *A. Gray*		S		M	A	N Y	L	K T
Preissianus, *Steetz*				A			L	K T G

Gnephosis, Cassini (1820).

Burkittii, *Bentham*			W		
eriocarpa, *Bentham*		C			
arachnoidea, *Turczaninow*		C	S		
cyathopappa, *Bentham* ...				M	
codonopappa, *F. v. Mueller*		C			
skirrophora, *Bentham* ...		C	S	M	

Calocephalus, R. Brown (1817).

Drummondii, *Bentham* ...			W	A		Y	L	T
Brownii, *F. v. Mueller* ...				A		Y	L	K G
Sonderi, *F. v. Mueller* ...		M						
lacteus, *Lessing* ...				A				
citreus, *Lessing* ...				A	N	Y		T G
platycephalus, *Bentham*	F	C	S					
Dittrichii, *F. v. Mueller*		C						

Eriochlamys, Sonder & F. v. Mueller (1852).

Behrii, *Sonder & F. v. Mueller*		M	N		L	K T
Knappii, *F. v. Mueller* ...	F					

Cephalipterum, A. Gray (1852).

Drummondii, *A. Gray* ...	W

Gnaphalodes, A. Gray (1852).

uliginosum, *A. Gray* ... ... S W M A N Y T

Pterocaulon, Elliot (1824).

sphacelatus, *Benth. & Hook.* ... F C S
Billardieri, *F. v. Mueller* ... F

Craspedia, G. Forster (1786).

Richea, *Cassini* ... ...A N Y L T G
chrysantha, *Bentham* ... ... C M A
globosa, *Bentham* ... ... S N
pleiocephala, *F. v. Mueller* ... C S W M

Chthonocephalus, Steetz (1845).

pseudevax, *Steetz* ... ... C W

Microseris, D. Don (1832).

Forsteri, *J. Hooker* ... ...A N Y L K T G

CANDOLLEACEAE.

Candollea, Labillardiere (1805).

graminifolia, *Swartz* ... ...A   K G
Tepperiana, *F. v. Mueller* ... ... K
calcarata, *R. Brown* ... ...A Y K G
perpusilla, *Hooker* ... ... G
floribunda, *R. Brown* ... ... F
despecta, *R. Brown* ... ... S A N L K T G

Leewenhoekia, R. Brown (1810).

dubia, *Sonder* ... ...A Y L K T G

CAMPANULACEAE.

Lobelia, Linne (1737).

rhombifolia, *De Vriese* ... ...A   K
microsperma, *F. v. Mueller* ... ...A Y L K
Browniana, *Roem. & Schultes* ... ... G
simplicicaulis, *R. Brown* ... ... G
heterophylla, *Labillardiere* ... F
purpurascens, *R. Brown* ... ... G
pedunculata, *R. Brown* ... ...A G
concolor, *R. Brown* ... ... M
platycalyx, *F. v. Mueller* ... ... K G
anceps, *Thunberg* ... ...A N L K T G
pratioides, *Bentham* ... ... T G
Benthami, *F. v. Mueller* ... C

LIST OF SPECIES. 245

Isotoma, R. Brown (1810).

petraea, *F. v. Mueller*	F	C	S	W		N			
scapigera, *G. Don*							L		
fluviatilis, *F. v. Mueller*								T	G

Wahlenbergia, Schrader (1814).

gracilis, *DeCandolle*	F	C	S	W	M	A	N	Y	L	K	T	G

GOODENIACEAE.
Brunonia, Smith (1809).

australis, *Smith*	F			A			T	G

Leschenaultia, R. Brown (1810).

divaricata, *F. v. Mueller*	F	C
striata, *F. v. Mueller*	F	

Dampiera, R. Brown (1810).

stricta, *R. Brown*								G
candicans, *F. v. Mueller*	F							
rosmarinifolia, *Schlechtendal*		W	M	A	N	Y	L	T
marifolia, *Bentham*			M					
lanceolata, *Cunningham*			M				K	

Velleya, Smith (1798).

paradoxa, *R. Brown*			M	A	N	Y	L	G
connata, *F. v. Mueller*	F	W	M					

Selliera, Cavanilles (1799).

radicans, *Cavanilles*	A	L	K

Catosperma, Bentham (1868).

Muelleri, *Bentham*	F

Scaevola, Linne (1771).

spinescens, *R. Brown*	F	C	S	W	M		Y		
Groeneri, *F. v. Mueller*				W					
crassifolia, *Labillardiere*				W		A	Y	L	K	G
parvifolia, *F. v. Mueller*	F								
depauperata, *R. Brown*	F	C							
collaris, *F. v. Mueller*		C		W					
suaveolens, *R. Brown*								T	G
microcarpa, *Cavanilles*						A	N		G
aemula, *R. Brown*	F		S	W				L	K	G
humilis, *R. Brown*			S	W			N		K
ovalifolia, *R. Brown*	F	C		W					
linearis, *R. Brown*						A	Y	L	K

Goodenia, Smith (1794).

Ramelii, *F. v. Mueller*	F								
humilis, *R. Brown*									G
amplexans, *F. v. Mueller*				A	N		K		
ovata, *Smith*				A		Y	L	K	G
varia, *R. Brown*		S	W	M	A	N	Y	L	K
Vilmoriniae, *F. v. Mueller*	F								
grandiflora, *Sims*	F								
Chambersii, *F. v. Mueller*	F								
albiflora, *Schlechtendal*				A	N	Y			
calcarata, *F. v. Mueller*		S	W						
Nicholsoni, *F. v. Mueller*	F								
Mitchellii, *Bentham*	F C								
heterochila, *F. v. Mueller*	F								
sepalosa, *F. v. Mueller*	F								
Mueckeana, *F. v. Mueller*	F								
Strangfordii, *F. v. Mueller*	F								
geniculata, *R. Brown*			W	M	A	N	Y	L	K T G
hirsuta, *F. v. Mueller*	F								
cycloptera, *R. Brown*	F	C	S	W			Y		
glauca, *F. v. Mueller*		C	S		M	A	N	Y	
microptera, *F. v. Mueller*		C							
elongata, *Labillardiere*									G
heteromera, *F. v. Mueller*		C		M					
pinnatifida, *Schlecht.*		S	W	M	A	N	Y	L	T G
pusilliflora, *F. v. Mueller*		S	W						

Calogyne, R. Brown (1810).

Berardiana, *F. v. Mueller* ... F

PRIMULACEAE.

Centunculus, Linne (1753).

minimus, *Linne* ... G

Samolus, Linne (1753).

repens, *Persoon* ... F   M A N Y L K T G

CONVOLVULACEAE.

Ipomoea, Linne (1737).

Davenporti, *F. v. Mueller* ... F
costata, *F. v. Mueller* ... F
Muelleri, *Bentham* ... F
heterophylla, *R. Brown* ... F C

Convolvulus, Linne (1753).

erubescens, *Sims* ... F C S W M A N Y L K T G
sepium, *Linne* ... A   T G

LIST OF SPECIES.

Polymeria, R. Brown (1810).
longifolia, *Lindley* ... ... F C
augusta, *F. v. Mueller* ... ... F C

Breweria, R. Brown (1810).
rosea, *F. v. Mueller* ... ... F
media, *R. Brown* C

Evolvulus, Linne (1763).
linifolius, *Linne* ... ... F C

Cressa, Linne (1747).
Cretica, *Linne* ... ... M

Dichondra, Forster (1776).
repens, *Forster* ... ... S   A N   L K T G

Wilsonia, R. Brown (1810).
humilis, *R. Brown* ... ...A   Y L K
rotundifolia, *Hooker* ... ...A   Y L K T
Backhousii, *Hooker* ... ... K T G

CUSCUTA, Linne.
australis, *R. Brown* ... ... F
Tasmanica, *Engelmann*... ... ... T

BORAGINEAE.
Coldenia, Linne (1747).
procumbens, *Linne* C

Heliotropium, Linne.
curassavicum, *Linne* ... ... F C S   M A   T
pleiopterum, *F. v. Mueller* ... F
Europaeum, *Linne* ... ... S   N
undulatum, *Vahl* ... ... F C   W
asperrimum, *R. Brown*... ... F C S   A N
ovalifolium, *Forskael* C
filaginoides, *Bentham* ... ... F C
tenuifolium, *R. Brown*... ... F

Halgania, Gaudichaud (1826).
cyanea, *Lindley* ... ... F C S W M A   Y L
lavandulacea, *Endlicher* ... ... W   A   Y   K T

Pollichia, Medikus (1783).
Zeylanica, *F. v. Mueller* ... F C S

Rochelia, Reichenbach (1824).

Maccoya, *F. v. Mueller* ... ... M

Echinospermum, Swartz.

concavum, *F. v. Mueller* ... C S W M A N Y

Eritrichium, Schrader (1820).

Australasicum, *DeCandolle* ... ... S M N Y G

Myosotis, Linne.

australis, *R. Brown* ... ... M A Y K G

Cynoglossum, Linne.

suaveolens, *R. Brown* ... ... M A N Y T G
australe, *R. Brown* ... ...A L T G
Drummondii, *Bentham*... ... F C S N

ASCLEPIADEAE.

Sarcostemma, R. Brown (1809).

australe, *R. Brown* ... ... F C S W M N

Cynanchum, Linne (1737).

floribundum, *R. Brown* ... F C S

Daemia, R. Brown (1809.)

Kempeana, *F. v. Mueller* ... F C

Marsdenia, R. Brown (1809).

Leichhardtiana, *F. v. Mueller* F C S M

APOCYNEAE.

Carissa, Linne (1767).

Brownii, *F. v. Mueller*... ... F

Alyxia, Banks (1810).

buxifolia, *R. Brown* ... ...A N Y L K T G

Notonerium, Bentham (1876).

Gossei, *Bentham* ... ... F

GENTIANEAE..

Sebaea, Solander (1810).

ovata, *R. Brown* ... ... M A N Y L K T G
albidiflora, *F. v. Mueller* ... ... G

LIST OF SPECIES.

Erythraea, Persoon (1805).
spicata, *Persoon* ... ... F C   A N Y L K T G

Gentiana, Linne.
saxosa, *Forster* ... ... T G

Limnanthemum, Gmelin.
crenatum, *F. v. Mueller* ... ... M
reniformis, *R. Brown* ... ... S M A N   L K T G

JASMINEAE.
Jasminum.
lineare, *R. Brown* ... ... F S   M   N
calcareum, *F. v. Mueller* ... F

PLANTAGINEAE.
Plantago, Linne.
varia, *R. Brown* ... ... C S W M A N Y L K T G

LOGANIACEAE.
Mitrasacme, Labillardiere (1804).
pilosa, *Labillardiere* ... ... G
paradoxa, *R. Brown* ... ... M A N Y L K   G
distylis, *F. v. Mueller* ... ... A   K

Logania, R. Brown (1810).
longifolia, *R. Brown* ... ... A N
crassifolia, *R. Brown* ... ... L K T G
ovata, *R. Brown* ... ... A   Y L K   G
stenophylla, *F. v. Mueller* ... ... W
linifolia, *Schlecht.* ... ... A   Y   T
nuda, *F. v. Mueller* ... ... W M

SOLANACEAE.
Solanum, Linne.
nigrum, *Linne* ... ... C S   A N   K   G
aviculare, *Forster* ... ... G
simile, *F. v. Mueller* ... ... S W M A N Y L K
fasciculatum, *F. v. Mueller* ... ... W   Y L
ferocissimum, *Lindley* ... ... F S
orbiculatum, *Dunal* ... ... C W
oligacanthum, *F. v. Mueller* ... F ?
esuriale, *Lindley* ... ... F C W M   N
chenopodium, *F. v. Mueller* ... C

250   FLORA OF SOUTH AUSTRALIA.

Sturtianum, *F. v. Mueller* ... F C S W
hystrix, *R. Brown* ... ... W
eremophilum, *F. v. Mueller* ... C S   N
lacunarium, *F. v. Mueller* ... C S W
petrophilum, *F. v. Mueller* ... F C S W
ellipticum, *R. Brown* ... ... F C S W

Lycium, Linne (1737).
australe, *F. v. Mueller* ... ... S W M A   Y

Anthotroche, Endlicher (1839).
Blackii, *F. v. Mueller* ... ... F

Datura, Linne (1737).
Leichhardtii, *F. v. Mueller* ... F S

Nicotiana, Linne.
suaveolens, *Lehmann* ... ... F C S W M A N Y L K   G

Duboisia, R. Brown (1810).
Hopwoodi, *F. v. Mueller* ... F   W

Anthocercis, Labillardiere (1806).
anisantha, *Endlicher* ... ... W   L
angustifolia, *F. v. Mueller* ... ... S   A
myosotidea, *F. v. Mueller* ... ... K T
Eadesii, *F. v. Mueller* ... ...   T

EPACRIDEAE.

Brachyloma, Sonder (1845).
ericoides, *Sonder* ... ... M A   K
daphnoides, *Bentham* ... ...   T
ciliatum, *Bentham* ... ...A   T G

Styphelia, Solander (1786).
adscendens, *R. Brown* ... ... G
pusilliflora, *F. v. Mueller* ... ... G
Sonderi, *F. v. Mueller* ... ... W   A   L K T G
humifusa, *Persoon* ... ...A N Y L K T G
strigosa, *Smith* ... ...A   L K
australis, *F. v. Mueller* ... ...A   G
Richei, *Labillardiere* ... ...A N Y L K T G
costata, *F. v. Mueller* ... ... K
striata, *Sprengel* ... ... W
collina, *Labillardiere* ... ...   T
hirsuta, *F. v. Mueller* ... ... K
concurva, *F. v. Mueller* ... ...A   K
virgata, *Labillardiere* ... ...A   T G
ericoides, *Smith* ... ... G

LIST OF SPECIES.

cordifolia, *F. v. Mueller*		W	M	A		Y	L	
hirtella, *F. v. Mueller*				A			K	
rufa, *F. v. Mueller*				A		L	K	
Woodsii, *F. v. Mueller*							K	G
serrulata, *Labillardiere*				A			K	
patula, *Sprengel*				A		Y L	K	
ovalifolia, *Sprengel*				A	N	Y L	K	T G
depressa, *Sprengel*				A			K	T
fasciculiflora, *F. v. Mueller*				A			K	
elliptica, *Smith*								G

Epacris, Cavanilles (1797).

impressa, *Labillardiere*			A		K T G
obtusifolia, *Smith*					G
lanuginosa, *Labillardiere*					G
microphylla, *R. Brown*					G

Sprengelia, Smith (1794).

incarnata, *Smith*			A	K G

LABIATAE.
Mentha, Linne.

australis, *R. Brown*	F C		M	A				T G
gracilis, *R. Brown*				A				T G
satureioides, *R. Brown*		S		A	N	Y L		T G

Teucrium, Linne.

sessiliflorum, *Bentham*			W	M	N	Y L	
integrifolium, *F. v. Mueller*	F						
corymbosum, *R. Brown*		S	W		N		G
racemosum, *R. Brown*	F C	S	W	M	A N	Y L	

Ajuga, Linne (1737).

australis, *R. Brown*		S	M	A N	K G

Microcorys, R. Brown (1810).

Macrediana, *F. v. Mueller*	F

Westringia, Smith (1797).

rigida, *R. Brown*		W	M	A N	Y L	K T
Dampieri, *R. Brown*		W			L	

Lycopus, Linne.

australis, *R. Brown*		M A		T G

Plectranthus, L'Heritier (1785).

parviflorus, *Willdenow*	F

Prunella, Linne.

vulgaris, *Linne*	...	...	...	...	...A					G

Scutellaria, Linne.

humilis, *R. Brown*	...	...	...	...	...	...	...K	

Prostanthera, Labillardiere (1806).

lasiantha, *Labillardiere*	...	...	...	...	...	...	...	...	G
rotundifolia, *R. Brown*	...	...	...	...	...	...	...	T	
striatiflora, *F. v. Mueller*	...	F	C	S	W				
Wilkieana, *F. v. Mueller*	...	F							
eurybioides, *F. v. Mueller*	...	...	...	...A					
spinosa, *F. v. Mueller*	...	...	S			L	K	T	
Behriana, *Schlechtendal*	...	...	...	...A				T	
Baxteri, *Cunningham*	...	...		W					
ringens, *Bentham*	...	...	F						
coccinea, *F. v. Mueller*	...	...	...	M	A	Y	L	K	T
chlorantha, *F. v. Mueller*	...	...	...	...A		Y	L	K	
calycina, *F. v. Mueller*	...	...	...	...	...	L			

LENTIBULARINEAE.
Utricularia, Linne (1737).

flexuosa, *Vahl*	...	...	...	...	...	...	...	...	G
dichotoma, *Labillardiere*	...	...	...	...A			K	T	G
lateriflora, *R. Brown*	...	...	...	...	...	...	...	...	G

Polypompholyx, Lehmann (1844).

tenella, *Lehmann*	...	...	...	...	...A			K	G

OROBANCHEAE.
Orobanche, Linne.

australiana, *F. v. Mueller*	...	C	S	W	A	Y	L

SCROPHULARINEAE.
Mimulus, Linne (1741).

gracilis, *R. Brown*	...	...	...	...	...	...	...	...	G
repens, *R. Brown*	...	...	...	...	M	A	N	L	K
prostratus, *Bentham*	...	...	F	C					

Mazus, Loureiro (1790).

pumilio, *R. Brown*	...	...	...	...	...	...	...	...	G

Buechnera, Linne (1737).

linearis, *R. Brown*	..	...F

LIST OF SPECIES.

Limosella, Linne.

aquatica, *Linne* ... ... S M A N L K G
Curdieana, *F. v. Mueller* ... ... S W M

Peplidium, Delile (1813).

humifusum, *Delile* ... ... F
Muelleri, *Bentham* C

Glossostigma, Arnott (1836).

Drummondii, *Bentham*... ... ... M
elatinoides, *Bentham* ... ... M

Euphrasia, Linne.

Brownii, *F. v. Mueller* ... ... W A N Y L K T G
scabra, *R. Brown* ... ...A L G

Stemodia, Linne (1759).

Morgania, *F. v. Mueller* ... F C S M A
viscosa, *Roxburgh* ... ... F
pedicellaris, *F. v. Mueller* ... F

Gratiola, Linne.

pedunculata, *R. Brown*... ... ... M
Peruviana, *Linne* ... ... M A K G

Veronica, Linne.

decorosa, *F. v. Mueller*... ... ... S N
Derwentia, *Andrews* ... ...A K G
gracilis, *R. Brown* ... ...A N G
distans, *R. Brown* ... ...A L K T
calycina, *R. Brown* ... ...Y K G
peregrina, *Linne* ... ...A N G

BIGNONIACEAE.

Tecoma, Jussieu (1789).

australis, *R. Brown* ... ... F

ACANTHACEAE.

Justicia, Linne (1737).

procumbens, *Linne* ... ... F C S
Bonneyana, *F. v. Mueller* ... ... M
Kempeana, *F. v. Mueller* ... ... F

Ruellia, Linne.

australis, *R. Brown* ... ... C M
primulacea, *F. v. Mueller* ... F C

FLORA OF SOUTH AUSTRALIA.

PEDALINEAE.
Josephinia, Ventenat (1804).
Eugeniae, *F. v. Mueller* ... C

VERBENACEAE.
Verbena, Linne.
officinalis, *Linne* ... ... M A T G
macrostachya, *F. v. Mueller* ... F C

Newcastlia, F. v. Mueller (1857).
cladotricha, *F. v. Mueller* ... F
spodiotricha, *F. v. Mueller* ... F C
cephalantha, *F. v. Mueller* ... F
bracteosa, *F. v. Mueller* ... F
Dixoni, *F. v. Mueller & Tate* ... ... M N

Dicrastylis, Drummond & Harvey (1855).
ochrotricha, *F. v. Mueller* ... F
Gilesii, *F. v. Mueller* ... ... F
Doranii, *F. v. Mueller* ... ... F
Beveridgei, *F. v. Mueller* ... F W
Lewellini, *F. v. Mueller* ... F

Clerodendrum, Burmann (1737).
floribundum, *R. Brown* ... ... F

Spartothamnus, Cunningham (1830).
teucriiflorus, *F. v. Mueller* ... F
puberulus, *F. v. Mueller* ... F

Avicennia, Linne (1737).
officinalis, *Linne* ... ... A N Y L

MYOPORINEAE.
Myoporum, Banks & Solander (1786).
montanum, *R. Brown* ... ... F C S M
insulare, *R. Brown* ... ... A N Y L K T G
viscosum, *R. Brown* ... ... A N Y K G
deserti, *Cunningham* ... ... S W M N Y
humile, *R. Brown* ... ... W M A Y L K G
brevipes, *Bentham* C
platycarpum, *R. Brown* ... ... S W M A N Y T

Eremophila, R. Brown (1810).
Dalyana, *F. v. Mueller* C
scoparia, *F. v. Mueller* ... ... C S W M N T
Delisserii, *F. v. Mueller* ... ... W

LIST OF SPECIES. 255

crassifolia, *F. v. Mueller*								L	
Behriana, *F. v. Mueller*					A	Y	L	K	
Weldii, *F. v. Mueller*			W						
Christophori, *F. v. Mueller*	F								
densifolia, *F. v. Mueller*									
gibbosifolia, *F. v. Mueller*					A		L		T
divaricata, *F. v. Mueller*				M					T
polyclada, *F. v. Mueller*		C		M					
Goodwinii, *F. v. Mueller*	F	C							
Elderi, *F. v. Mueller*	F		S						
Willsii, *F. v. Mueller*	F								
santalina, *F. v Mueller*			S						
longifolia, *F. v. Mueller*	F	C	S	W	M		N	Y	
Freelingii, *F. v. Mueller*	F	C	S						
bignoniflora, *F. v. Mueller*	F	C			M				T
MacDonneli, *F. v. Mueller*	F	C	S	W					
Bowmani, *F. v. Mueller*		C							
rotundifolia, *F. v. Mueller*	F	C							
leucophylla, *Bentham*	F								
Paisleyi, *F. v. Mueller*				W					
Sturtii, *R. Brown*	F	C							
exilifolia, *F. v. Mueller*	F								
Mitchelli, *Bentham*	F								
Gibsoni, *F. v. Mueller*	F								
Berryi, *F. v. Mueller*				W					
Clarkei, *F. v. Mueller*	F								
Gilesii, *F. v. Mueller*	F	C							
Hughesii, *F v. Mueller*	F								
oppositifolia, *R. Brown*			S	W			N		
Latrobei, *F. v. Mueller*	F	C	S	W					
Brownii, *F. v. Mueller*	F	C	S	W	M			Y	K T
Duttonii, *F. v. Mueller*		C	S						
Maculata, *F. v. Mueller*	F	C	S	W	M				
denticulata, *F. v. Mueller*				W					
latifolia, *F. v. Mueller*		C	S	W	M				
alternifolia, *R. Brown*	F		S	W	M		N		

CONIFERAE.

Callitris, Ventenat (1808).

verrucosa, *R. Brown*	F		S		M	A	N	Y	L K T
cupressiformis, *Ventenat*						A			K

CYCADEAE.

Encephalartos, Lehmann, (1834).

MacDonnelli, *F. v. Mueller* ... F

HYDROCHARIDEAE.
Ottelia, Persoon (1805).
ovalifolia, *Richard* ... ... M A   K   G

Vallisneria, Linne (1753).
spiralis, *Linne* ... ... M   G

Blyxa, Noronha (1806).
Roxburghii, *Richard* C

Hydrilla, Richard (1811).
verticillata, *Caspary* ... ... M

Halophila, DuPetit-Thouars (1806).
ovalis, *Hooker* ... ... A   Y L K   G

ORCHIDEAE.
Dipodium, R. Brown (1810).
punctatum, *R. Brown* ... ... A   G

Cymbidium, Swartz (1799).
canaliculatum, *R. Brown* ... C

Thelymitra, Forster (1776)
ixioides, *Swartz* ... ... A   G
longifolia, *Forster* ... ... A N Y   K   G
parviflora, *R. Brown* ... ... A
aristata, *Lindley* ... ... A   L   G
grandiflora, *Fitzgerald* ... ... A
fuscolutea, *R. Brown* ... ... A   L
luteocilium, *Fitzgerald* ... ... A
urnalis, *Fitzgerald* ... ... A
flexuosa, *Endlicher* ... ... A   K
antennifera, *Hooker* ... ... A   Y   K T G
carnea, *R. Brown* ... ... A   Y
rubra, *Fitzgerald* ... ... A   G

Calochilus, R. Brown (1810).
Robertsoni, *Bentham* ... ... A   G

Diuris, Smith (1798).
punctata, *Smith* ... ... G
palustris, *Lindley* ... ... A   Y L   T G
maculata, *Smith* ... ...
pedunculata, *R. Brown* ... ... A   Y   G
sulphurea, *R. Brown* ... ... A   K G
longifolia, *R. Brown* ... ... A   L K G

LIST OF SPECIES.

Orthoceras, R. Brown (1810).

strictum, *R. Brown*	...	...	...	...	A				

Cryptostylis, R. Brown (1810).

longifolia, *R. Brown*	...	...	...	...	...	...			G

Prasophyllum, R. Brown (1810).

elatum, *R. Brown*	...	...	...	...	A	Y		K	G
australe, *R. Brown*	...	...	...	...			...		G
fuscum, *R. Brown*	...	...	...	...	A	Y	L		
patens, *R. Brown*	...	...	...	...	A	Y			G
despectans, *Hooker*	...	...	...	...	A				
nigricans, *R. Brown*	...	...	...	...			L		

Spiranthes, L. C. Richard (1818).

australis, *Lindley*	...	...	...	...	...	...			G

Microtis, R. Brown (1810).

porrifolia, *R. Brown*	...	...	...	A	N	Y	L	K T	G
minutiflora, *F. v. Mueller*	...	...	...	A					G

Corysanthes, R. Brown (1810).

pruinosa, *Cunningham*	...	...	...	...	A			K	G

Pterostylis, R. Brown (1810).

concinna, *R. Brown*	...	...	...	...	A				
nana, *R. Brown*	...	...	...	W				K	G
nutans, *R. Brown*	...	...	...	...	...			K	G
pedunculata, *R. Brown*	...	...	...	...	A				
curta, *R. Brown*	...	...	...	...	A				
cucullata, *R. Brown*	...	...	...	...	A				G
praecox, *R. Brown*	...	...	...	...	...			K	
reflexa, *R. Brown*	...	...	...	...	A	Y			
obtusa, *R. Brown*	...	...	...	...					
barbata, *Lindley*	...	...	...	...	A	Y	L	K	
mutica, *R. Brown*	...	...	...	W		Y	L		G
rufa, *R. Brown*	...	...	S	W	A	Y			
longifolia, *R. Brown*	...	...	...	...	A		L		G
vittata, *Lindley*	...	...	...	...	A	Y		K	

Acianthus, R. Brown (1810).

caudatus, *R. Brown*	...	...	...	...		...		K	
exsertus, *R. Brown*	...	...	...	...	A	Y			

Cyrtostylis, R. Brown (1810).

reniformis, *R. Brown*	...	...	...	A		Y	L	K	G

R

258   FLORA OF SOUTH AUSTRALIA.

Glossodia, R. Brown (1810).

major, *R. Brown* ... ...A   Y   G

Lyperanthus, R. Brown (1810).

nigricans, *R. Brown* ... ...A   Y   K

Eriochilus, R. Brown (1810).

autumnalis *R. Brown* ... ...A   Y   K   G
fimbriatus, *F. v. Mueller* ... ...A

Caladenia, R. Brown (1810).

Cairnsiana, *F. v. Mueller* ... ...A
reticulata, *Fitzgerald* ... ...A
toxochila, *Tate* ... ... W
tentaculata, *Tate* ... ... W
Menziesii, *R. Brown* ... ...A   G
filamentosa, *R. Brown* ... ... K   G
dilatata, *R. Brown* ... ...A   Y   L   K   T   G
Patersoni, *R. Brown* ... ...A N Y L   G
leptochila, *Fitzgerald* ... ...A N
latifolia, *R. Brown* ... ...A   Y   L   K   G
coerulea, *R. Brown* ... ...A
carnea, *R. Brown* ... ...A   Y   K   G
deformis, *R. Brown* ... ...A   Y   L   K   T   G

IRIDEAE.

Patersonia, R. Brown (1807).

glauca, *R. Brown* ... ...A   K   G
longiscapa, *Sweet* ... ...A   T   G

Sisyrinchium, Linne (1737).

cyaneum, *Lindley* ... ...A   K   G

AMARYLLIDEAE.

Hypoxis, Linne (1759).

hygrometrica, *Labillardiere* ... ...   G
glabella, *R. Brown* ... ...A   L   T   G
pusilla, *Hooker* ... ... W M A N Y L

Crinum, Linne (1737).

angustifolium *R. Brown* ... F
flaccidum, *Herbert* ... ... C S
pedunculatum, *R. Brown* ... ... M

LIST OF SPECIES.

Calostemma, R. Brown (1810).

purpureum, *R. Brown*					A	N	Y			
luteum, *Sims*				C		M				

LILACEAE.

Burchardia, R. Brown (1810).

umbellata, *R. Brown*					A		Y	L	K	G

Wurmbea, Thunberg (1781).

dioica, *F. v. Mueller*		F C S		M	A	N	Y	L	K	G

Xerotes, R. Brown (1810).

longifolia, *R. Brown*					A				T G
dura, *F. v. Mueller*		S			A	N	Y		
Brownii, *F. v. Mueller*					A				G
effusa, *Lindley*			W	M	A		Y		G
micrantha, *Endlich.*					A			L	
Thunbergii, *F. v. Mueller*					A				
glauca, *R. Brown*			W		A		Y	L	T G
elongata, *Bentham*									G
leucocephala, *R. Brown*	F		W	M	A		Y		
juncea, *F. v. Mueller*					A			L	T

Dianella, Lamarck (1780).

revoluta, *R. Brown*			W	M	A	N	Y		G
laevis, *R. Brown*					A			K	G

Bulbine, Linne (1737).

bulbosa, *Haworth*					A	N	Y	L		G
semibarbata, *Haworth*		C S W	M	A	N	Y	L	K	T G

Caesia, R. Brown (1810).

vittata, *R. Brown*				A	Y	G
parviflora, *R. Brown*				A		

Arthropodium, R. Brown (1810).

paniculatum, *R. Brown*								G
minus, *R. Brown*						Y	L	
strictum, *R. Brown*		S		A	N	Y	K	T G
fimbriatum, *R. Brown*				A			L K	

Thysanotus, R. Brown (1810).

dichotomus, *R. Brown*	F			A			L K	G
tuberosus, *R. Brown*	F	S		A			K	T G
exasperatus, *F. v. Mueller*		S				Y		
Baueri, *R. Brown*			W	M	A		Y	G
Patersoni, *R. Brown*					A	N	Y	L K T G
exiliflorus, *F. v. Mueller*	F C							

FLORA OF SOUTH AUSTRALIA.

Tricoryne, R. Brown (1810).
elatior, *R. Brown* ... ... C A Y L G

Chamaescilla, F. v. Mueller (1870).
corymbosa, *F. v. Mueller* ... ... A Y L K G

Laxmannia, R. Brown (1810).
sessiliflora, *Decaisne* ... ... A K T G

Calectasia, R. Brown (1810).
cyanea, *R. Brown* ... ... T G

Corynotheca, F. v. Mueller (1870).
lateriflora, *F. v. Mueller* ... F M

Xanthorrhoea, Smith (1798).
quadrangulata, *F. v. Mueller* ... ... A K T G
Tateana, *F. v. Mueller* ... ... A L K
minor, *R. Brown* ... ... G
semiplana, *F. v. Mueller* ... ... S A N L T G

XYRIDEAE.
Xyris, Linne (1737).
operculata, *Labillardiere* ... ... A
gracilis, *R. Brown* ... ... G

COMMELINEAE.
Commelina, Linne.
ensifolia, *R. Brown* ... F

ALISMACEAE.
Damasonium, Jussieu.
australe, *Salisbury* ... ... M A

JUNCACEAE.
Luzula, DeCandolle (1805).
campestris, *DeCandolle* ... ... A N L K G

Juncus, Linne.
planifolius, *R. Brown* ... C A T G
caespititius, *Meyer* ... ... A N K G
bufonius, *Linne* ... ... S M A N Y L K T G

LIST OF SPECIES.

homalocaulis, *F. v. Mueller* ... G
pallidus, *R. Brown* ... S M A N Y L K T G
pauciflorus, *R. Brown* ... A K G
communis, *Meyer* ... S W M A N Y L T G
prismatocarpus, *R. Brown* ... S M A N G
maritimus, *Lamarck* ... A L K

PALMAE.
Livistona, R. Brown (1810).

Mariae, *F. v. Mueller* ... F

TYPHACEAE.
Typha, Linne.

angustifolia, *Linne* ... F S M A N L K T G

FLUVIALES.
Triglochin, Linne.

centrocarpa, *Hooker* ... C S W M A Y K G
mucronata, *R. Brown* ... A L K T G
striata, *Ruiz & Pavon* ... M A N L K T G
procera, *R. Brown* ... M A K G

Potamogeton, Linne.

Tepperi, *Bennett* ... S M A N L K T G
tennicaulis, *F. v. Mueller* ...
crispus, *Linne* ... M
ochreatus, *Raoul* ... A K
acutifolius, *Link* ... M
pectinatus, *Linne* ... A L K

Posidonia, Koenig (1806).

australis, *Hooker* ... A Y

Ruppia, Linne (1735).

maritima, *Linne* ... A L K T G

Zostera, Linne (1747).

nana, *Mertens* ... A Y G
Tasmanica, *Mertens* ... A

Lepilaena, Drummond (1855).

Preissii, *F. v. Mueller* ... C A L K T G
australis, *Drummond* ... T

FLORA OF SOUTH AUSTRALIA.

Cymodocea, Koenig (1806).
antarctica, *Endlicher* ... ... A   Y   L

Naias, Linne (1735).
tenuifolia, *R Brown* ... ... M
major, *Allioni* ... ... F

LEMNACEAE.
Lemna, Linna (1735).
trisulca, *Linne* ... ... M   K   G
minor, *Linne* ... ... M   K   G
oligorrhiza, *Kurz* ... ... T

Wolffia, Horkel (1839).
Michelii, *Schleiden* ... ... M   G

RESTIACEAE.
Trithuria, J. Hooker (1860).
submersa, *Hooker* ... ... M   K

Aphelia, R. Brown (1810).
gracilis, *Sonder* ... ... A   K   G
pumilio, *F. v. Mueller* ... ... A   K   G

Centrolepis, Labillardiere (1804).
polygyna, *Hieron* ... ... M   L K   G
glabra, *F. v. Mueller* ... ... M
aristata, *Roem. & Schult.* ... ... S   M A   Y L K T G
fascicularis, *Labillardiere* ... ... A   G
strigosa, *Roem. & Schult.* ... ... S   A N   L K   G

Lepyrodia, R. Brown (1810).
Muelleri, *Bentham* ... ... G

Restio, Linne (1767).
complanatus, *R. Brown* ... ... G
tetraphyllus, *Labillardiere* ... ... G

Leptocarpus, R. Brown (1810).
tenax, *R. Brown* ... ... A   K   G
Brownii, *Hooker* ... ... A   Y L   G

Calostrophus, Labillardiere (1806).
lateriflorus, *F. v. Mueller* ... ... A   G
fastigiatus, *F. v. Mueller* ... ... A   K   G

Lepidobolus, Nees (1846).

drapetocoleus, *F. v. Mueller* ... ...A G

CYPERACEAE.
Lepidospora, F. v. Mueller (1875).

tenuissima, *F. v. Mueller* ... ...A G

Kyllingia, Rottboell (1773).

monocephala, *Rottboell*... ... ... M
intermedia, *R. Brown* ... ...A

Cyperus, Linne.

eragrostis, *Vahl* ...		C		A				G
tenellus, *Linne* ...				...A				
gracilis, *R. Brown*		C						
pygmaeus, *Rottboell*		C						
squarrosus, *Linne*		C						
difformis, *Linne*...		C		M				
trinervis, *R. Brown*				M				
vaginatus, *R. Brown*	F	C	S	M	A	N	L	T
holoschoenus, *R. Brown*	F							
Gilesii, *Bentham*		C						
fulvus, *R. Brown*		C						
alterniflorus, *R. Brown*...			S	W				
Iria, *Linne*		C						
diphyllus, *Retzius*		C						
subulatus, *R. Brown*	F	C						
rotundus, *Linne*...	F	C						
lucidus, *R. Brown*				M	A			
exaltatus, *Retzius*			S	M				

Schoenus, Linne (1737).

capillaris, *F. v. Mueller*							K	G
aphyllus, *Boeckeler*			M					
brevifolius, *R. Brown* ...			...A					G
apogon, *Röm. & Schult.*			...A	N	Y	L	K	G
axillaris, *Poiret* ...			...A					
sculptus, *Boeckeler*						L	K	
fluitans, *Hooker*...							K	
sphaerocephalus, *Poiret*								G
nitens, *Poiret*			...A		Y	L	K	T G
deformis, *Poiret*...						L		
Tepperi, *F. v. Mueller* ...		W	A		Y	L	K	
discifer, *Tate*							K	

Fimbristylis, Vahl (1806).

communis, *Kunth*			S	M				
velata, *R. Brown*		C		M				
ferruginea, *Vahl*	F	C						
barbata, *Bentham*		C						
Neilsoni, *F. v. Mueller*				M				

Heleocharis, R. Brown (1810).

sphacelata, *R. Brown*		S	A	N		K	G
acuta, *R. Brown*		S	M A	N	L	K T	G
multicaulis, *Smith*		S	A	N		K	G
acicularis, *R. Brown*			M				

Scirpus, Linne.

pungens, *Vahl*			M	A	N	L	G
maritimus, *Linne*			M	A			
lacustris, *Linne*	C	S	M	A	N		G
litoralis, *Schrader*	F	S		A	N		

Isolepis, R. Brown.

fluitans, *R. Brown*			A			K	G
setacea, *R. Brown*			A			K	
riparia, *R. Brown*			M A	N	Y L	K	G
cartilaginea, *R. Brown*	C	S	M A	N	Y L	K	G
inundata, *R. Brown*		S	M A	N		K	G
supina, *R. Brown*			M				
nodosa, *R. Brown*		S	M A	N	Y L	K	G

Lipocarpha, R. Brown (1818).

monocephala, *R. Brown* ... ... M

Fuirena, Rottboell (1818).

glomerata, *Lamarck* ... ... F

Carex, Linne.

inversa, *R. Brown*			A				G
chlorantha, *R. Brown*			A				
tereticaulis, *F. v. Mueller*		S	A	N	L K T	
paniculata, *Linne*			A			K	G
caespitosa, *R. Brown*		S	A	N			G
pumila, *Thunberg*			A	N			G
breviculmis, *R. Brown*			A				G
Gunniana, *Boott*		S	A				G
pseudocyperus, *Linne*			A			K	G

Caustis, R. Brown (1810).

pentandra, *R. Brown*			A		K	G

LIST OF SPECIES.

Cladium, P. Browne (1756).

mariscus, *R. Brown*				...A				G
articulatum, *R. Brown*			S	A				G
glomeratum, *R. Brown*				...A				G
tetraquetrum, *Hooker*				...A			K	
schoenoides, *R. Brown*				...A		L	K	T G
filum, *R. Brown*				...A	Y	L	K	G
Gunnii, *Hooker*				...A				
junceum, *R. Brown*			S	A	N	L	K	G
trifidum, *F. v. Mueller*				...A				
radula, *R. Brown*				...A				G
psittacorum, *F v. Mueller*				...A				G
lanigerum, *R. Brown*				...A	Y	L		
deustum, *R. Brown*				...A	Y	L	K	T

Lepidosperma, Labillardiere (1804).

longitudinale, *Labillardiere*								G
exaltatum, *R. Brown*								G
gladiatum, *Labillardiere*				...A	Y	L	K	T G
elatius, *Labillardiere*				...A				
concavum, *R. Brown*				...A				
viscidum, *R. Brown*			S W	A	Y	L	K	G
laterale, *R. Brown*				M A				G
congestum, *R. Brown*						L		
globosum, *Labillardiere*								G
lineare, *R. Brown*				...A			K	
semiteres, *F. v. Mueller*				...A				T
canescens, *Boeck.*				...A				T G
filiforme, *Labillardiere*				...A			K	T G
carphoides, *F. v. Mueller*				...A		L	K	T G

Chorizandra, R. Brown (1810).

enodis, *Nees*				...A	Y	L	K	G

GRAMINEAE.

Setaria, Palisot (1812).

glauca, *Palisot*		F
macrostachya, *H. B. & K.*		F
viridis, *Palisot*		F C S

Pennisetum, L. Richard (1805).

refractum, *F. v. Mueller*		F C

Panicum, Linne.

coenicolum, *F. v. Mueller*		C S	M	
divaricatissimum, *R. Brown*	F	S	M A	
prolutum, *F. v. Mueller*			M A	N

FLORA OF SOUTH AUSTRALIA.

Species	F	C	S	W	M	A	N	Y	L	K	T	G
effusum, *R. Brown*		C	S	W	M	A	N					G
Mitchelli, *Bentham*		C										
decompositum, *R. Brown*	F	C	S	W								
spinescens, *R. Brown*					M							
Crus-galli, *Linne*					M	A						
adspersum, *Trinius*		C										
pauciflorum, *R Brown*	F											
leucophaeum, *H. B. K.*	F	C	S	W			N					
argenteum, *R. Brown*	F											
gracile, *R. Brown*			S		M	A						
helopus, *Trinius*		C	S		M							
Gilesii, *Bentham*	F	C										
distachyum, *Linne*	F	C										
reversum, *F. v. Mueller*	F	C	S									

Spinifex, Linne (1767).

Species	F	C	S	W	M	A	N	Y	L	K	T	G
paradoxus, *Bentham*	F	C	S									
hirsutus, *Labillardiere*						A		Y	L	K	T	G

Neurachne, R. Brown (1810).

Species	F	C	S	W	M	A	N	Y	L	K	T	G
alopecuroides, *R. Brown*	F					A	N	Y	L		T	G
Mitchelliana, *Nees*					M							
Munroi, *F. v. Mueller*				W	M							

Hemarthria, R. Brown (1810).

Species	F	C	S	W	M	A	N	Y	L	K	T	G
compressa, *R. Brown*							N	A				

Imperata, Cyrillo (1792).

Species	F	C	S	W	M	A	N	Y	L	K	T	G
arundinacea, *Cyrillo*	F					A			L			

Erianthus, L. Richard (1803).

Species	F	C	S	W	M	A	N	Y	L	K	T	G
fulvus, *Kunth*	F	C			M							

Andropogon, Linne.

Species	F	C	S	W	M	A	N	Y	L	K	T	G
sericeus, *R. Brown*	F	C	S				N					
pertusus, *Willdenow*	F											
annulatus, *Forskael*		C										
punctatus, *Roxburgh*		C	S				N					
exaltatus, *R. Brown*	F	C	S			A	N					
bombycinus, *R. Brown*	F				M	A					T	
gryllus, *Linne*	F											

Anthistiria, Linne, fil. (1779).

Species	F	C	S	W	M	A	N	Y	L	K	T	G
ciliata, *Linne*	F	C	S			A	N	Y	L		T	G
avenacea, *F. v. Mueller*	F	C			M							
membranacea, *Lindley*		C										

LIST OF SPECIES.

Eriochloa, Humboldt & Kunth (1815).
polystachya, *Humboldt & Kunth*   F C S   M

Perotis, Aiton (1789).
rara, *R. Brown* ...   ... F C S

Tragus, Haller (1768).
racemosus, *Haller*   ... F C S W M

Ehrharta, Thunberg (1779).
stipoides, *Labillardiere*... ... ...A   K   G

Pappophorum, Schreber (1791).
commune, *F. v. Mueller* ... F C S W M A N

Lepturus, R. Brown (1810).
incurvatus, *Trinius* ... ... M A N Y L K T
cylindricus, *Trinius* ... ...A   G

Echinopogon, Palisot (1812).
ovatus, *Palisot* ... ...A N   K   G

Alopecurus, Linne (1735).
geniculatus, *Linne* ... ... S   M A   G

Stipa, Linne (1737).
elegantissima, *Labillardiere* ... ... S W M A   Y
Tuckeri, *F. v. Mueller*... ... ... M
teretifolia, *Steudel* ... ... Y L K
flavescens, *Labillardiere* ... ...A   L
Muelleri, *Tate* ... ...A
setacea, *R. Brown* ... ... S   M   N   L
semibarbata, *R. Brown*... ... C   A   Y L K T G
pubescens, *R. Brown* ... ... S   M   N
aristiglumis, *F. v. Mueller* ... ... S   M   N Y L K
scabra, *Lindley* ... ... F C S   M A   Y   G

Chloris, Swartz (1788).
pectinata, *Bentham* C
acicularis, *Lindley* ... ... F C S   M   N
truncata, *R. Brown* ... ... F   S
barbata, *Swartz*... ... ... F C
scariosa, *F. v. Mueller* ... ... F

Dichelachne, Endlicher (1833).
crinita, *J. Hooker* ... ... S   A N Y   K   G
sciurea, *Hooker* ... ...A   K   G

Agrostis, Linne (1735).

scabra, *Willdenow*								G
venusta, *Trinius*								G
densa, *F. v. Mueller*				A				
Solandri, *F. v. Mueller*				M A	N	Y L	K T	G
quadriseta, *R. Brown*				A	N		K	
montana, *R. Brown*				A				

Aristida, Linne (1753).

stipoides, *R. Brown*	F C	S				
arenaria, *R. Brown*	F C	S	M A			
leptopoda, *Bentham*			M			
Behriana, *F. v. Mueller*			M	A N Y		T
ramosa, *R. Brown*	F	S				
calycina, *R. Brown*	F	S				

Amphipogon, R. Brown (1810).

strictus, *R. Brown*	W	A	Y	G

Pentapogon, R. Brown (1810).

Billardieri, *R. Brown*	A

Sporobolus, R. Brown (1810).

Virginicus, *Humboldt & Kunth*	C S	M A		Y L K	G
Indicus, *R. Brown*	F				
Lindleyi, *Bentham*	F C S W				
actinocladus, *F. v. Mueller*	F C W				

Cynodon, L. Richard (1805).

Dactylon, *Richard*		M A	G
convergens, *F. v. Mueller*	C		
ciliaris, *Bentham*	C		

Aira, Linne (1737).

caespitosa, *Linne*	G

Eriachne, R. Brown (1810).

aristidea, *F. v. Mueller*	C	
ovata, *Nees*	C	
pallida, *F. v. Mueller*	F	
scleranthoides, *F. v. Mueller*	F	W
mucronata, *R. Brown*	F	
obtusa, *R. Brown*		M

Triraphis, R. Brown (1810).

mollis, *R. Brown*	F C S W M

LIST OF SPECIES.

Danthonia, DeCandolle (1805).

bipartita, *F. v. Mueller*	...	...	...	M						
carphoides, *F. v. Mueller*	...	...	...	M	N					
penicillata, *F. v. Mueller*	...	...	S	W	A	N	Y	L	K	G
nervosa, *Hooker*	...	...	S	M	A	N	Y		K	G

Astrebla, F. v. Mueller (1876).

pectinata, *F. v. Mueller*	...	F	C	S
triticoides, *F. v. Mueller*	...		C	

Agropyron, Gaertner (1770).

scabrum, *Palisot*	...	...	...	...	M	A	N	Y	K T G

Elytrophorus, Palisot (1812).

articulatus, *Palisot*	...	...	C		M

Arundo, Linne.

Phragmites, *Linne*	...	F		M	A	N	L	T G

Bromus, Linne (1735).

arenarius, *Labillardiere*	...	F	S	W	M	A	N	Y	L	K T G

Festuca, Linne.

duriuscula, *Linne*	...	...	...	...	...	A N

Diplachne, Palisot (1812).

loliiformis, *F. v. Mueller*	...	F	C	S
Muelleri, *Bentham*	...	...	C	
fusca, *Palisot*	...	...	...	C	M

Schedonorus, Palisot (1812).

litoralis, *Palisot*	...	...	...	S W	M	A	N	L K T

Distichlis, Rafinesque (1819).

maritima, *Rafinesque*	...	...	...	...	...	A N Y	K T G

Eleusine, Gaertner (1788).

cruciata, *Lamarck*	...	...	F	C	S	M
digitata, *Sprengel*	...	...	F			

Triodia, R. Brown (1810).

Mitchelli, *Bentham*	...	...	F					
pungens, *R. Brown*	...	...	F					
irritans, *R. Brown*	...	...	F	C	S	W	M	A N Y

Eragrostis, Palisot (1812).

tenella, *Palisot*	F	C							
trichophylla, *Bentham*		C	S	W					
leptocarpa, *Bentham*		C	S						
pilosa, *Palisot*					M				G
diandra, *Steudel*					M	A			
Brownii, *Nees*	F		S	W	M	A	L	T	G
concinna, *Steudel*	F	C							
speciosa, *Steudel*		C							
laniflora, *Bentham*		C	S						
chaetophylla, *Steudel*		C	S		M				
eriopoda, *Bentham*					M				
lacunaria, *F. v. Mueller*					M				
falcata, *Gaudichaud*		C			M				

Poa, Linne (1737).

Billardieri, *Steudel*									G
nodosa, *Nees*					M	A	N	L	T G
caespitosa, *Forster*			S	W		A	N Y	L K	T G
lepida, *F. v. Mueller*				W	M		N Y	L K	T
fluitans, *Scopoli*						A			
Fordeana *F. v. Mueller*					M				
syrtica, *F. v. Mueller*						A	Y	K	T
ramigera, *F. v. Mueller*		C	S		M				

LYCOPODIACEAE.

Lycopodium, Linne.

Carolinianum, *Linne*	A		
laterale, *R. Brown*	A	K	
densum, *Labillardiere*	A		

Selaginella, Palisot (1805).

Preissiana, *Spring*	A	K	T	G
uliginosa, *Spring*				G

RHIZOSPERMAE.

Azolla, Lamarck (1783).

pinnata, *R. Brown*	M	
filiculoides, *Lamarck*	M	G

Marsilea, Linne (1735).

quadrifolia, *Linne*	F	C	S	M	A	G

Pilularia, Linne.

globulifera, *Linne*	S

LIST OF SPECIES.

FILICES.

Ophioglossum, Linne.

vulgatum, *Linne* ... ... C S W M A   Y   K   G

Botrychium, Swartz (1800).

ternatum, *Swartz* ... ... A

Schizaea, Smith (1791).

fistulosa, *Labillardiere* ... ... A   K
bifida, *Swartz* ... ... G

Gleichenia, Smith (1791).

circinata, *Swartz* ... ... A

Osmunda, Linne.

barbara, *Thunberg* ... ... A

Dicksonia, L'Heritier (1788).

Billardieri, *F. v. Mueller* ... ... A   G

Lindsaea, Dryander (1791).

linearis, *Swartz* ... ... A   G

Adiantum, Linne.

Æthiopicum, *Linne* ... ... W   A   K   G

Pteris, Linne (1735).

aquilina, *Linne* ... ... A N   L K   G
arguta, *Aiton*   L   G
incisa, *Thunberg* ... ... G

Lomaria, Willdenow (1809).

discolor, *Willdenow* ... ... A   K
lanceolata, *Sprengel* ... ... G
Capensis, *Willdenow* ... ... A   K

Asplenium, Linne (1737).

flabellifolium, *Cavanilles* ... ... A N   L   G
furcatum, *Thunberg* ... ... G
bulbiferum, *Forster* ... ... G

Aspidium, Swartz (1800).

molle, *Swartz* ... ... M		
decompositum, *Sprengel* ... ... G		

Polypodium, Linne.

punctatum, *Thunberg* ... ... G

Grammitis, Swartz (1800).

Reynoldsii, *F. v. Mueller*	...	F							
rutaefolia, *R. Brown*	... ...	F	C	S	W	A	N	Y	
leptophylla, *Swartz*	... ...	...A					N		K

Cheilanthes, Swartz (1806).

tenuifolia, *Swartz*	... ...	F	C	S	W	A	N	Y	L	K	G
vellea, *F. v. Mueller*	... ...	F	C	S	W					
distans, *Braun*					W	A	N	Y		
Clelandi, *F. v. Mueller & Tate*	...				W					

———o———

This FLORA includes :—

Orders 101, Genera 553, Species 1,935.

EXPLANATION OF SPECIES-NAMES

(PERSONAL NAMES OMITTED).

RULES FOR PRONOUNCIATION.

In classical names there are as many syllables as there are vowels, even if terminal, except in the case of diphthongs and when *u* with any vowel follows *g*, *q* or *s*.

A unaccented, ending a word, is pronounced like *ah*.

I unaccented, if final, is sounded as if written *eye;* and when it ends a syllable not final it has the sound of *e*, as *Behr-e-eye* for *Behrii*.

C is pronounced like *k* before *a*, *o* and *u*; but is soft before *e*, *i* and *y*.

G is pronounced hard before *a*, *o* and *u*; soft like *J* before *e*, *i* and *y*.

T, *s* and *c* before *ia*, *ie*, *ii* and *eu*, when preceded by the accent, change their sound, *t* into *tsh*, *s* and *c* into *sh* or *zh*; but when the accent is on the first diphthong the preceding consonant preserves its sound, as *aurantiaca*

Ch before a vowel is pronounced like *k*.

Cn, *gn*, *ps*, *pt* and other uncombinable consonants, when they begin a word, the first letter is not sounded; in the middle of a word they are separate.

Ph are pronounced like *f*.

Sch sounds like *sk*.

S at the end of a word has a hissing sound; except when preceded by *e*, *r* or *n*, when it sounds like *z*.

X at the beginning of a word sounds like *z*.

The accented syllable is indicated by the mark (′) at the end, as in *acero′sa*.

acanthoc′lada; thorn-branched
acero′sa; needle-shape
achilleoï′des; Achillea-like
acicula′ris; needle-like
acid′ula; somewhat acid
acina′cea; dagger-like
acrade′nia; having a gland at the tip
acrop′tera; summit-winged
actinoc′ladus; ray-branched
aculea′ta, um; prickly

acumina′ta, um; long-pointed
acu′ta; sharp-pointed
acutan′gula; sharp-cornered
acutifo′lius; having pointed leaves
aden′ophorum; gland-bearing
adpres′sa; pressed close to ——
adscen′dens; ascending
adsper′sum; spotted
æ′mula; rivalling
æquilatera′le, equal-sided
Æthio′picum; Ethiopian

s

agrifo'lia; having sharp leaves
al'ba; white
albidiflo'ra; with whitish flowers
albiflo'ra; white-flowered
alopecuroi'des; } Alopecurus-like
alopecuroi'deus;
alterniflo'rus; alternate-flowered
alternifo'lia; alternate-leaved
ama'ra; bitter
ambig'uum; doubtful
America'na; American
ammannioi'des; Ammannia-like
ammoch'aris; sand-loving
ammoph'ilum; sand-loving
amphib'ium; amphibious
amplex'ans; clasping
amygda'lina; almond-like
an'ceps; two-edged
Andic'ola; Andes-dwelling
anethifo'lius; Anethum-leaved
aneu'ra; veinless
angula'ta, um; angular
angus'ta; narrow
angustifo'lius, a, um; narrow-leaved
anisan'tha; unequal-flowered
annula'tus; ringed
anom'alum; unusual
anseri'na; of a goose
Antarc'tica; Antarctic
antennif'era; antennae-bearing
anthemoi'des; Anthemum-like
apet'ala; without petals
aphyl'lus, a; leafless
apicula'tum; distinctly pointed
apo'gon; without a beard
aquat'ica, aquat'ilis; living in water
aquili'na; crooked like an eagle's beak
aquifo'lium; holly-leaved
arachnoi'dea; cobweb-like
arbus'cula; somewhat shrubby
arcua'ta; bow-shaped, arched
arena'ria; belonging to sand
argen'teum; silvery
argu'ta; pretty
arista'ta; awned
aristi'dea; Aristida-like
aristiglu'mis; having awned glumes

arma'ta; armed
artemisioi'des; Artemisia-like
arthrop'oda; joint-stalked
articula'ta, um; jointed
arundina'cea; reed-like
Asiat'ica; Asiatic
as'pera; rough
asper'rimum; very rough
asterot'richus; star-haired
astrocar'pus; star-fruited
atriplic'inum; Atriplex-like
atropurpu'rea; black-purple
attenua'tum; becoming slender
au'rea; golden
auric'omum; golden-haired
auricula'ta; having ear-like lobes
Australas'ica, um; Australasian
australia'na; southern
austra'lis, e; southern
autumna'lis; flowering in autumn
avena'cea; oat-like
axilla'ris; in the axils
baccif'era; berry-bearing
Balonnen'sis; from the R. Balonne, Queensland
balsam'ica; balsam-like
barba'ta; bearded
bar'bara; foreign
basal'tica; growing on basalt
betonicifo'lia; Betonica-leaved
bi'color; two-coloured
bicorn'is; two-horned
bicuspida'ta; two-speared
bicus'pis; with two spears
bi'dens; with two teeth
bif'ida, um; two-cleft
biflo'ra; two-flowered
bignonia'ceus; Bignonia-like
bignoniflo'ra; Bignonia-flowered
bina'ta; by couples
biparti'ta; nearly divided in two parts
bipinna'ta; twice pinnate
Blenno'dia; a generic name
bombyc'inus; made of silk
boronifo'lia; Boronia-leaved
brachia'tus; branched
brachyan'drus; having short stamens

EXPLANATION OF SPECIES-NAMES.

brachybot′rya ; having short bunches or racemes
brachyglos′sus ; short-tongued
brachypap′pus ; having a short pappus
brachyphyl′lum ; short-leaved
brachyp′tera ; short-winged
brachysipho′nius ; short-tubed
bracteo′sa ; having bracts
brevicul′mis ; short-stemed
brev′idens ; having short teeth
brevifo′lius, a ; short-leaved
brevipeda′ta ; short-stalked
brev′ipes ; having short stalks
bufo′nius ; belonging to a toad
bulbo′sa ; bulbous
bulbif′erum ; bulb-bearing
bursarifo′lia ; Bursaria-leaved
buxifo′lia ; Buxus (box)-leaved
cæru′lea ; blue
cærules′cens ; bluish
cæspit′itius ; turfy
cæspito′sa ; turfy
calamifo′lia ; reed-leaved
calcara′ta ; spurred
calca′reus, um ; pertaining to lime-stone
callicar′pa, calocar′pa ; beautiful-fruited
calyc′inus, a, um ; having a prominent calyx
calym′ega ; with large calyx
calyptra′ta ; capped or covered
calyxhyme′nia ; with a membranous calyx
campes′tris ; belonging to fields
campylan′tha, bent-flowered
canalicula′ta, um ; channelled
can′dicans ; whitish
canes′cens ; greyish, hoary
cannabi′na ; hemp-like
Capen′sis ; of the Cape of Good Hope
capilla′ris ; hair-like
capita′tus ; headed
capitella′ta ; small-headed
cardaminoï′des ; Cardamine-like
cardiocar′pa ; having a heart-shaped fruit
cardiophyl′la ; heart-leaved

carina′tum ; keeled
car′nea ; flesh-coloured
carphoï′des ; Carpha-like
cartilag′ineus ; gristly
cassinia′cea ; Cassinia-like
cauda′tus ; tailed
celastroï′des ; Celastrus-like
centrocar′pa ; having spurred fruits
cephalan′tha ; having flowers in heads
ceratophyl′lus ; with horned leaves
chætophyl′la ; bristle-leaved
chenopo′dium ; goose-foot
Chinen′sis ; of China
chloran′tha ; green-flowered
chordophyl′la ; with string-like leaves
chrysan′tha ; golden-flowered
chrysoglos′sa ; golden-tongued
ciba′ria ; yielding food
cilia′ris ; cilia′ta, um ; having cilia, or eye-lashes
cineras′cens ; ashy in colour
cine′rea, um ; ash-coloured
circina′ta ; flat-coiled
cit′reus ; citron-coloured
clandesti′na ; hidden
clavellifo′lia ; with little knob-like leaves
cneorifo′lia ; knife-leaved
coactilifo′lia ; woolly-leaved
coccin′eus, a ; scarlet
cochleari′na ; Cochlearia-like
cochlear′is ; coiled like a snail-shell
codonapap′pa ; with a bell-shaped pappus
cœnic′olum ; dirt-dwelling
colla′ris ; necklaced
colletioï′des ; Colletia-like
colli′na ; of a hill
colutoï′des ; Coluta-like
commu′nis, e ; common
comocar′pa ; having hair-tufted fruits
complana′tus ; smoothed
compres′sa ; flattened
con′cavum ; concave
concin′na ; neat

con'color; of one colour
concur'va, um; bent towards one another
congest'um; crowded
conna'ta; grown together
conoceph'ala; cone-headed
contin'ua; joined without interruption
converg'ens; leading to one point
cordifo'lia; with heart-shaped leaves
coria'cea; leathery
cornicula'ta; small-horned
corolla'ta; having a corolla
coronillifo'lia; Coronilla-leaved
coronopifo'lia; Coronopus-leaved
corrigiola'cea; Corrigiola-like
corruga'ta; furrowed
corymbiflo'rum; having flowers in corymbs
corymbo'sa, um; having corymbs
corynoc'alyx; having a club-shaped calyx
cosmophyl'la; regular-leaved
costa'ta; ribbed
cotinifo'lius; Cotinus-leaved
Cot'ula; a generic name
craspedocar'pa; fringe-fruited
crassifo'lia; thick-leaved
crassius'cula; somewhat thick
crena'tum; round-notched
Cre'tica; belonging to Crete
crini'ta; long-haired
cris'pus; curled
crista'tum; crested
cro'ceum; yellow
crucia'ta; crossed
Crus-gal'li; "the leg of a fowl"
cryphiopet'ala, cryptopet'alum; having hidden petals
crystal'linus, a, um; crystalline
cuculla'ta; hooded
cuneifo'lia,; wedge-leaved
cupressifor'mis; Cypress-like
Curassa'vicum; belonging to Curaçoa
cur'ta; short
curviflo'ra; curve-flowered
cur'vipes; curve-stalked
cyan'ea, um; dark-blue

cyanopet'alus; having blue petals
cyathopap'pa; having a cup-shaped pappus
cyclo'pis; circle-eyed
cyclop'tera; circle-winged
cygno'rum; of the swans, from Swan River, W. Aust.
cylindri'cus, a; cylindrical
cymbacan'tha; boat-flowered
cymo'sa; cyme-bearing
cyperophyl'la; Cyperus-leaved
dac'tylon; a finger
daphnoï'des; Daphne-like
dealba'ta; whitened
de'bilis; weak
decan'dra; with 10 stamens
decap'tera; ten-winged
decip'iens; deceptive
decompos'itum; having various compound divisions
decoro'sa; graceful
decum'bens; lying down
decur'rens; running down
decussa'ta; crossed
deform'is; ill-shapen
demer'sum; under water
den'sa, um; dense
densifo'lia; dense-leaved
dentatifo'lius; having toothed leaves
denticula'ta; small toothed
denuda'ta; naked
depaupera'ta; impoverished
depres'sa; flattened down
Derwen'tia; of the R. Derwent, Tasmania
deser'ti; of the desert
desola'ta; desolate, as regards habitat
despec'tans; despising
despec'tum; despised
deus'ta; burnt (appearance)
diacan'tha; with two thorns
dian'der, ra, rum; with two stamens
dian'tha; two-flowered
dichot'omus, a; repeatedly forked
dictyoph'leba; net-veined
diffor'mis; ill-shapen
diffu'sa; spread out

EXPLANATION OF SPECIES-NAMES. 277

digita'ta ; fingered, parts radiating
dig'yna ; with two pistils
dimorphol'epis; having two forms of scales
dioi'ca ; double-housed, sexes in distinct plants
dipterocar'pa ; having two-winged fruits
diphyl'lus : two-leaved
dis'cifer ; disk-bearing
dis'color ; of a different colour
dissec'ta ; much cut
dissitiflo'ra ; with scattered flowers
distach'yum ; double-spiked, or in two rows
dis'tans ; wide apart
disty'lis, a ; having two styles
divarica'ta ; spreading widely
divaricatis'simum ; most divaricate
diversifo'lia ; various-leaved
dodonæifo'lia ; Dodonæa-leaved
doratox'ylon ; spear-wood
drapetoco'leus ; instable-sheathed
dry'adeus : a mythological name
du'bia ; doubtful
du'ra ; hard
durius'cula ; somewhat hard
echinops'ila ; Echinops-like
effu'sa, um ; poured out
ege'na ; in want of
elachan'tha ; small-flowered
elachis'tum ; very small
ela'ta ; tall
elatinoï'des ; Elatine-like
ela'tior ; taller
ela'tius, um ; lofty
elegantis'sima ; most elegant
ellip'tica, um ; acutely oval
elonga'tus ; lengthened
empetrifo'lia ; Empetrum-leaved
Emphy'sopus ; swollen foot
enchylænoï'des ; Enchylæna-like
enneaphyl'la ; nine-leaved
enneasper'mus ; nine-seeded
eno'dis ; without knots
ensifo'lia ; sword-leaved
Eragros'tis ; a generic name
eremoph'ila, um; desert-loving
erian'tha ; woolly-flowered

erica'cea ; heath-like
ericifo'lia ; heath-leaved
ericoï'des ; Erica (heath) -like
erina'cea ; hedgehog-like, prickly
eriocar'pus, a ; woolly fruited
erioch'iton ; having a woolly coat
eriop'oda ; woolly stalked
erubes'cens ; somewhat red
erythran'tha ; red-flowered
estrophiola'ta ; (seed) without a " strophiole "
Europæ'um ; European
eurybioï'des ; Eurybia-like
eusty'lis ; with a well developed style
exalta'tus, um ; raised
exaspera'tus ; much roughened
exiguifo'lius, a ; thin-leaved
exig'uum ; thin
exiliflo'rus ; slender-flowered
exilifo'lia ; slender-leaved
exi'lis ; slender
Exocar'pi ; upon Exocarpos
exocarpoï'des ; Exocarpos-like
expan'sa ; spread out
exser'tus ; thrust out
ex'ul ; an exile
falca'ta ; sickle-shaped
farino'sa ; mealy
Farnesia'na ; Farnesian
fascicula'ris ; in bundles
fascicula'ta, um ; bundled
fasciculiflo'ra ; having flowers in bundles
fastigia'tus ; pointed at the top
ferocis'simum ; most fierce
ferrugin'ea, um ; rusty
filaginoï'des ; Filago-like
filiculoï'des ; Filicula-like
filifo'lia, um ; thread-leaved
filifor'mis, e ; thread-like
fi'lum ; a cord
filamento'sa ; full of threads
fimbria'tum ; fringed
fimbriola'ta ; somewhat fringed
fissival've ; split-valved
fistulo'sa ; pipe-like
flabellifo'lium ; fan-leaved
flac'cidum ; weak
fla'va ; yellow

flaves'cens ; yellowish
flaviflo'ra ; yellow-flowered
flexuo'sus, a ; bending
floribun'dus, a, um; rich in flowers
flu'itans ; floating
fluvia'tilis ; belonging to rivers
folio'sum ; leafy
fruticulo'sum ; somewhat shrubby
furca'tum ; forked
fusco-lu'tea ; brown-yellow
fus'cum, a ; brown, tawny
gamophyl'la ; with united leaves
geminifo'lia ; twin-leaved
genetylloi'des ; Genetyllis-like
genicula'tus, a ; bent at the joint
genistifo'lia ; Genista-leaved
genistioi'des ; Genista-like
gibber'ulus ; somewhat humped
gibbo'sa ; swollen, humped
gibbosifo'lia ; having swellings on the leaves
glabel'la ; somewhat glabrous
glaber'rima ; most glabrous
gla'bra ; without hairs
gladia'tum ; sword-shaped
glandulig'era ; glandule-bearing
glandulo'sus ; very glandular
glau'ca ; sea-green
glauces'cens; glaucous-like
glaucifo'lia ; glaucous-leaved
globo'sus, um ; globular
globulif'era ; little globe-bearing
glomera'tus, a, um ; heaped together
glutino'sa ; sticky
glutes'cens ; somewhat sticky
gomphrenoi'des ; Gomphrena-like
gonioc'alyx ; having an angular calyx
goniocar'pa; having angular fruits
goniophyl'la ; angle-leaved
grac'ilis, e ; slender
gramin'ea; grass-like
graminifo'lia ; grass-leaved
grandibract'eus ; having large bracts
grandiflo'ra, um ; large flowered
grave'olens ; offensive smelling
Gryl'lus ; a cricket (a plant eaten by)

hakeæfo'lius ; Hakea-leaved
hakeoi'des ; Hakea-like
halimoi'des; Halimus-like
halmaturi'na ; inhabiting Kangaroo Island
haloph'ilum; sea-loving
hedera'cea ; ivy-like
helichrysoi'des ; Helichrysum-like
helipteroi'des ; Helipterum-like
he'lopus ; twist-footed
hemiglau'ca ; half-glaucous
hemiphloi'a ; half-barked
hemistei'rus ; half-barren
heteran'thum ; irregular-flowered
heterochi'la ; irregular-lipped
heterom'era ; having variable parts
heterophyl'la ; irregular-leaved
hexan'dra, um ; with 6 stamens
hirsu'tus, a ; hairy
hir'ta; hairy
hirtel'la ; slightly hairy
his'pida ; rough with stiff hairs
hispid'ula ; somewhat hispid
holocar'pum ; entire-fruited
holoschœ'nus ; entirely like Schœnus
homalocau'lis ; equal-stalked
homalophyl'la ; equal-leaved
hor'rida ; terrible
humifu'sa, um ; spread on the ground
humil'lima ; most lowly
hu'milis ; lowly
humistra'ta ; spread on the ground
hyalosper'mum ; glass-seeded
hydrop'iper ; water-pepper
hygrome'trica ; sensitive to moisture
hypoleu'cus; underside white
hyssopifo lia ; Hyssop-leaved
hys'trix ; (prickly as) a porcupine
ilicifo'lia ; holly-leaved
implexic'oma ; having entwined foliage
impres'sa ; stamped
inca'nus, a ; hoary
incarna'ta ; flesh-coloured
inci'sa ; jagged
inclu'sa ; enclosed
incrassa'ta, um ; thickened

EXPLANATION OF SPECIES-NAMES. 279

incurva'tus; bent in
In'dicus, a, um; Indian
indu'tum; clad
insula're; on islands
integer'rima; most entire
integrifo'lia, um; entire-leaved
interme'dius, a, um; betwixt
intrica'ta; entangled
inunda'tus; (subject to be) over-flowered
inver'sa; turned over
involucra'ta; having an involucre
iodocar'pum; violet-fruited
I'ria; proper name
irri'tans; provoking
iteaphyl'la; willow-leaved
ixioï'des; Ixia-like
Japon'icum; Japanese
jun'cea, um; rush-like
juncifo'lia; rush-leaved
juniper'ina; juniper-like
Ka'li; yielding salt
labicheoï'des; Labichea-like
lacinia'ta; jagged
lac'tens; milky
lacuna'ria, um; belonging to lagoons
lacus'tris; belonging to lakes
læ'vis, e; smooth
læviga'tum; smoothened
lan'ata; woolly
lanceola'ta; spear-shaped
lanicus'pis; woolly speared
laniflo'ra; woolly flowered
lanig'era, um: wool-bearing
lano'sa; woolly
lanugino'sa; downy
lapathifo'lium; "Dock"-leaved
lappa'ceus; having burs
lappula'cea; somewhat bur-like
largiflo'rens; with a broad inflorescence
lasian'tha; woolly flowered
lasiocar'pum; woolly fruited
latera'le; on one side
lateriflo'ra; having flowers on one side
latifo'lius, a, um; broad-leaved
lau'tus; washed
lavandula'cea; "Lavender"-like

Lawrencel'la; a generic name
lax'a; loose
laxiflo'ra; loose-flowered
leiostach'ya; smooth-spiked
lep'ida; scaly
lepidophloi'a; scale-barked
lepidophyl'lus; scale-leaved
lepido'tus; scaly
leptocar'pum; thin-fruited
leptol'epis; thin-scaled
leptopet'alum; having thin petals
leptophyl'la; thin-leaved
leptop'oda; thin-stalked
lessertifo'lia; Lessertia-leaved
leucan'tha; white-flowered
leucoceph'ala; white-headed
leucoc'oma; white-haired
leucopet'alum; having white petals
leucophæ'um; gray
leucophrac'ta; white-enclosed
leucophyl'la; white-leaved
leucopsid'ium: of a white appearance
leucop'tera; white-winged
leuc'oxylon; white wood
liatroï'des; Liatrus-like
ligus'trina; Privet-like
limba'tum; bordered
linarifo'lia; Linaria-leaved
linearifo'lius; narrow-leaved
linea'ris, e; narrow
linea'ta; streaked
linifo'lius; flax-leaved
linophyl'lus; flax-leaved
litora'lis; of the shore
lobiflo'ra; lobe-flowered
lobula'ta: having small lobes
loliiform'is; Lolium-like
longiflo'ra; long-flowered
longifo'lia; long-leaved
longis'capa; long-stalked
longitudina'le; lengthways
lo'rea; thong-like
lotifo'lia; Lotus-leaved
loxophyl'la; oblique-leaved
lu'cidum; shining
luteiflo'ra; yellow-flowered
luteo-al'bum; yellowish-white
lu'teum; yellow
lycopodifo'lia; Lycopodium-leaved

lysiphloï'a; smooth-barked
macran'tha; large-flowered
macrocar'pus, a; large-fruited
macroceph'alus; large-headed
macrorrhyn'cha; large-beaked
macrostach'ya; large-spiked
macroz'yga; long-yoked
ma'crum; thin
macula'ta; spotted
Maderaspata'na; belonging to Madras
magnif'icus; magnificent
magniflo'rus, a; large-flowered
ma'jor; larger
margina'le; at the margin
margina'ta; bordered
marifo'lia;
mari'na; of the sea
Maris'cus; "bull-rush"
marit'imus, a; belonging to the sea
medicagin'ea, medicaginoï'des; Medicago-like
me'dius; intermediate
megaglos'sus; great-tongued
megalodon'tus; great-toothed
megalop'tera; great-winged
meionect'es; rather small
melan'tha; black-flowered
melanocar'pa; black-fruited
melanox'ylon; blackwood
membrana'cea; membranous
mesembrian'themum; mid-day flowering
micran'thus, a; small-flowered
microcar'pa; small-fruited
microchlæ'nus; minutely clothed
microglos'sus; small-tongued
micropet'ala; small-petaled
microphyl'lus, a, um; small-leaved
microp'tera; small-winged
microsper'ma; small-seeded
microthe'ca; small-chambered (-fruited)
microz'yga; short-yoked
min'imus, a, um; smallest
mi'nor; smaller
mi'nus; small
minutiflo'ra; minute-flowered
minutifo'lia; minute-leaved

minu'tula; rather minute
mix'ta; mingled
mol'lis, e; soft
mollis'sima; very soft
monoceph'ala; one-headed
monophyl'la; one-leaved
monoplocoï'des; Monop'loca-like
monosper'ma; one-seeded
monta'na, um; belonging to mountains
Morga'nia; a generic name
moscha'tum; musky
mucrona'ta; short-pointed
multicau'lis; many-stemed
multiflo'ra; many-flowered
multisec'ta; much divided
multistria'ta; many-streaked
murica'ta; prickly
mu'tica; beardless
myosoti'dea; Myosotis-like
myrsinoï'des; Myrsine-like
myrtifo'lia; myrtle-leaved
myrtilloï'des; Myrtillus-like
na'na; dwarf
nasturtioï'des; Nasturtium-like
nematophyl'la; with scattered leaves
nemoro'sum; shade-dwelling
nephrosper'ma; kidney-seeded
nig'ricans; blackish
ni'grum; black
ni'tens; shining
nit'ida; neat
nitraria'ceum; Nitraria-like
no'bilis; remarkable
nodo'sus; knotty
nota'bilis; notable
nummula'rium; coin-like
nu'tans; nodding
obcorda'ta; reverse-cordate
obionoï'des; Obione-like
obli'qua; leaning to one side
obova'tus, um; reverse-ovate
ob'tusa; blunt
obtusan'gulum; blunt-angled
obtusifo'lius, a, um; blunt-leaved
ochrea'tus; having sheathing stipules
ochro'tricha; pale yellow-haired
octophyl'la; eight-leaved

EXPLANATION OF SPECIES-NAMES.

odontocar'pa; tooth-fruited
odora'tus, a; perfumed
officina'lis; used as a drug
oleifo'lium; olive-leaved
olera'cea; potherb-like
oleo'sa; oily
oligacan'thum; few-thorned
oligan'tha; few-flowered
oligophyl'la; few-leaved
opa'ca; dark
opercula'ta; having a lid
oppositiflo'ra; opposite-flowered
oppositifo'lia; opposite-leaved
orbicula'ris, orbicula'tum; of a round form
orienta'lis; eastern
orna'ta; ornamented
oroboï'des; Orobus-like
orygioï'des; Orygia-like
otocar'pum; ear-fruited
oxycar'pum; point-fruited
Oxyced'rus; a generic name
ovalifo'lia, um; oval-leaved
ova'lis, ova'ta; egg-shaped
ovi'na; belonging to sheep
pachyphyl'la; thick-leaved
pachyp'tera; thick-winged
pal'lidus, a; pale-coloured
paludo'sum, palus'tris; belonging to marshes
panicula'ta, um; having panicles
panno'sus; ragged
papillo'sum; covered with little pimples
papyrocar'pa; paper-fruited
parabol'ica; parabola-shaped
paradox'us, a; strange
par'va; small
parviflo'rus, a; small-flowered
parvifo'lius; small-leaved
pat'ens; exposed, spreading
pat'ula; spreading out
pauciflo'rus, a, um; few-flowered
pectina'tus, a; combed
pedicella'ris; having stalklets
peduncula'ta, um; stalked
pelta'ta; having a shield
pen'dulus; hanging
penicilla'ta; hair-tufted
pentan'dra, um; with 5 stamens

pentap'tera; five-winged
pentat'ropis; five-keeled
peregri'na; a wanderer
peren'nis; living more than two years
perpusil'lus, a; very little
persica'rium; peach-like
pertu'sus; perforated
Peruvia'na; Peruvian
petaloc'alyx; having a petal-like calyx
petiola'ris; having petioles
petræ'a; growing on rocks
petroph'ila, um; rock-loving
Peuce; a pine
phacoï'des; Phaca-like
phillyræoï'des; Phillyra-like
phlebopet'alum; vein-petaled
phlebophyl'la; vein-leaved
phragmi'tes; a reed
phylicoï'des; Phylicia-like
phyllodin'ea; having phyllods
picridioï'des; Picris-like
pilos'ula; somewhat hairy
pilo'sus, a, um; hairy
pimeloï'des; Pimelea-like
pinifo'lius; pine-leaved
pinna'ta; feathered
pinnatif'ida; feather-cleft
planifo'lius; flat-leaved
plantaginel'la; a little plantain (Plantago)
plantagin'eum; Plantago-like
platyc'alyx; having a broad calyx
platycar'pum; broad-fruited
platyceph'alus; broad-headed
platyglos'sa; broad-tongued
platyl'epis; broad-scaled
platyp'oda; broad-stalked
platyp'terus; broad-winged
plebe'ia, um; common
pleioceph'ala; many-headed
pleiococ'ca; many-fruited
pleiopet'ala; many-petaled
pleiop'terum; many-winged
pleurandroï'des; Pleuranda-like
pleurocar'pa; side-fruited
pleuropap'pus; having a pappus on one side
plumulif'era; plume-bearing

T

podolepid'eum ; stalk-scaled
polyan'dra ; with many stamens
polyc'lada ; many-branched
polygalifo'lia, um ; Polygala-leaved
polygaloi'des ; Polygala-like
polygonoi'des ; Polygonum-like
polyg'yna ; with many pistils
polystach'ya ; many-spiked
polystemo'nea ; with many stamens
polyz'yga ; many-paired
pomif'era ; apple-bearing
porphyroglos'sa ; purple-tongued
porrifo'lia ; leak-leaved
præ'cox ; early
pratioi'des ; Pratia-like
prenanthoi'des ; Prenanthus-like
primula'cea ; primrose-like
prismatocar'pus ; prism-fruited
prismatothe'cum ; prism-fruited
proce'ra ; tall
procum'bens ; bending down
prolu'tum ; washed
propin'qua ; related to
prore'pens ; creeping forward
prostra'ta, um ; lying flat
pruino'sa ; frosted
pseude'vax ; the false-Evax (a generic name)
pseudo-cype'rus ; the false-Cyperus
psittaco'rum ; of the parrots
psoralcoi'des ; Psoralea-like
pterocar'pa ; wing-fruited
pterochæ'tum ; wing-bristled
pterosper'ma ; wing-seeded
ptychosper'ma ; fold-seeded
puber'ulus ; somewhat downy
pubes'cens ; downy
pulchel'lus ; pretty
pu'mila ; dwarfish
pu'milio ; a dwarf
Pu'milo ; a generic name
puncta'ta, um ; dotted
punctula'ta ; somewhat dotted
pun'gens ; pricking
purpuras'cens ; purplish
purpura'ta ; clad in purple
purpu'rea ; purple
pusilliflo'ra ; small-flowered
pusil'lus, a ; small
pustula'ta ; covered with blisters

pycnan'tha ; dense-flowered
pygmæ'a ; dwarf
pyramida'lis, pyramida'ta ; pyramid-like
pyrifo'lia ; pear-leaved
pyrifor'mis ; pear-shaped
quadrangula'ta ; four-angled
quadridenta'ta ; four-toothed
quadrifa'ria ; four ways
quadrif'idum ; four-cleft
quadrifo'lia ; four-leaved
quadriparti'ta ; four-divided
quadrise'ta ; four-bristled
quadrival'vis ; four-valved
Quan'dang ; an aboriginal name
quinquecus'pis ; five-speared
racemig'era ; raceme-bearing
racemo'sa ; full of clusters
radi'cans ; rooting
ra'dula ; a scraper
ramig'era ; branch-bearing
ramo'sa ; full of branches
ramulo'sa ; full of branchlets
ra'ra ; scarce
recur'vus ; curved back
reflex'a ; bent back
refract'um ; broken
renifor'mis ; kidney-shaped
repan'da ; broad, flat
rep'ens ; creeping
rep'tans ; creeping along
retino'des ; net-like
retiven'ea ; net-veined
retu'sa, um ; blunt (with the tip turned down)
rever'sum ; turned upside down
rhadinost'achyum ; slender-spiked
rhagodioi'des ; Rhagodia-like
rhigiophyl'la ; shivering-leaved
rhizoceph'alus ; root-headed
rhombifo'lia ; rhombus-leaved
rhytidosper'mus ; wrinkle-seeded
rig'ens ; stiff
ripa'rius, a ; belonging to river-banks
rivula'ris ; belonging to rivulets
robus'ta ; robust, stout
ros'eum ; rose-coloured
rosmarinifo'lia ; rosemary-leaved
rostra'ta, um ; beaked

EXPLANATION OF SPECIES-NAMES.

rotundifo′lia ; round-leaved
rotun′dus, um ; round
rubioi′des ; Rubus-like
ru′bra ; red
rudera′le ; belonging to waste places
ru′fa ; reddish
ruga′ta ; wrinkled
rugo′sa ; wrinkled
rupic′ola ; rock-dwelling
rutæfo′lia ; rue-leaved
rutidoch′lamys ; having a wrinkled outer covering
rutidol′epis ; wrinkle-scaled
Salica′ria ; a generic name
salicifo′lia ; willow-leaved
salic′ina ; willow-like
salig′nus ; willowy
salsugino′sa ; full of salt-juice
sanguisor′bæ ; Sanguisorba-like
santalifo′lia ; Santalum-leaved
santal′ina ; Santalum-like
satureioi′des ; Sature′ia-like
saxo′sa ; living on stony ground
scabiosifo′lia ; Scabiosa-leaved
sca′bra ; rough
sca′brida ; somewhat rough
scan′dens ; climbing
scapig′era ; stalk-bearing
scario′sa ; having *dry* bracts
schœnoi′des ; Schœnus-like
scirpifo′lia ; Scirpus-leaved
sciu′rea ; squirrel-tailed
scleranthoi′des ; Scleran′thus-like
sclerophyl′la ; hard-leaved
scopa′ria, um ; broom-like
scorpioi′des ; curled over like a scorpion's tail
sculp′tus ; carved
sedifio′rus ; Se′dum-flowered
sedifo′lia ; Se′dum-leaved
semibacca′tum ; somewhat berried
semibarba′ta ; half-bearded
semifer′tile ; half-fruitful
semipappo′sum ; half-bearded
semipla′na ; half-smooth
semit′eres ; almost cylindrical
sen′tis ; a prickle
sepalo′sa ; full of sepals
se′pium ; a cuttle fish-bone

septif′raga ; breaking partitions
seri′ceus, a ; silky
serpyllifo′lia ; Serpyl′lum-leaved
serrula′ta ; minutely saw-edged
sessil′iceps ; sessile-headed
sessiliflo′ra, um ; sessile-flowered
seta′ceus, a ; bristly
seto′sa ; full of bristles
sidoi′des ; Si′da-like
Siemsse′nia ; a generic name
simil′e ; like, similar
sim′ulans ; resembling
sim′plex ; simple
simplicicau′lis ; simple-stemed
skirroph′ora ; hard rind-bearing
Soph′era ; an Arabic name
spar′tea ; broom-like
spartioi′des ; Spartium (broom)-like
spathula′tus, a, um ; spoon-shaped
specio′sa ; handsome
spectab′ilis ; worth seeing
Sper′gula ; "Spurrey-wort"
sphacela′tus, a ; withered, decayed
sphærocar′pa, um ; round-fruited
sphæroceph′alus ; round-headed
sphæros′pora ; round-seeded
spica′tus, a, um ; having spikes of flowers
spines′cens ; thorny
spino′sa ; thorny
spira′lis ; spiral
spodiot′richa ; ashy haired
spondylophyl′la ; spindle-leaved
spongiocar′pa ; spongy fruited
squama′tus ; scaly
squa′mea, scaly
squarro′sus, a ; very rough
stellig′era ; starry
stellula′tus ; full of little stars
stenobot′ryus, a ; narrow-bunched
stenophyl′lus, a ; narrow-leaved
stenoz′yga ; narrow-yoked
stipita′tum ; stalked
stipoi′des ; Sti′pa-like
stipula′ris, stipulig′era ; having stipules
stria′ta ; streaked
striatiflo′ra ; streak-flowered
stric′tus, a, um ; rigid

strigo'sa ; rough with stiff hairs
strongylophyl'la, um ; round-leaved
suædifo'lia ; Suæda-leaved
suave'olens ; sweet-smelling
suavis'sima ; most sweet
suber'osa ; corky
sublana'ta ; almost woolly
submer'sa ; sunk under water
subochrea'ta, with somewhat sheathing stipules
subula'tus ; awl-shaped
sulca'ta ; furrowed
sulphu'rea ; sulphur-coloured
supi'nus, a ; lifted up
sylves'tre ; growing in woods
synan'dra ; having united stamens
syr'tica ; belonging to quicksands
tabac'ina ; Tabaco (tobacco)-like
Tasman'ica ; Tasmanian
ten'ax ; tough
tenel'lus, a, um ; delicate
tentacula'ta ; having feelers
tenuicau'lis ; slender-stemed
tenuiflo'ra ; slender-flowered
tenuifo'lius, a, um ; slender-leaved
ten'uis ; slender, thin
tenuis'sima ; most slender
tereticau'lis ; cylindrical-stemed
teretifo'lius, a ; cylindrical-leaved
termina'lis ; ending
terna'tum ; three-grouped
terres'tris, e ; on the land
tessella'ris ; formed in chequers
tetrag'ona ; four-angled
tetragonophyl'la ; tetragonal-leaved
tetragy'na ; having four pistils
tetran'thera ; four-flowered
tetraphyl'lus, um ; four-leaved
tetraque'trum; having four sharp angles
teucriiflo'rus; Teucrium-flowered
teucrioï'des ; Teucrium-like
thesioï'des ; Thesium-like
thespidioï'des ; Thespidium-like
thymoï'des ; "thyme"-like
tillæa'cea ; Tillæa-like
tomento'sus, a ; covered with dense short hairs or tomentum

toxochi'la ; bow-lipped
trachycar'pa ; rough-fruited
trachysper'mus ; rough-seeded
trian'dra ; having 3 stamens
triangula're ; three-cornered
trichost'achya ; hairy-spiked
tricor'nis ; three-horned
trigonocar'pa; three-angled-fruited
trif'ida ; three-cleft
trifoliola'tum ; having three leaflets
trilocula'ris ; three-celled
triner'vis ; three-nerved
trineu'ra ; three-nerved
trio'num ; belonging to the North
triparti'ta ; three-divided
trip'tera ; three-winged
trisect'um ; three-cut
trisul'ca ; three-pointed
triticoï'des ; "wheat"-like
trunca'ta ; cut off
tubero'sus ; having tubers
tubuliflo'rus ; tubular-flowered
tubulo'sa, um ; pipe-like
ulic'ina ; "furze"- or "gorse"-like
uligino'sa ; living in moist places
umbella'ta ; umbel-bearing
umbro'sum ; living in shade
uncina'ta ; hooked at the end
undula'tum ; wavy
uniflo'ra ; one-flowered
unifoliola'ta ; with one leaflet
urba'num ; belonging to a town
urna'lis ; urn-like
vagina'tus ; sheathed
va'ria ; changeable
vela'ta ; covered
vel'lea ; woolly
velutinel'lum ; somewhat velvety
veluti'num ; velvety
ven'usta ; graceful
verbesinoï'des ; Verbesina-like
ver'na ; flowering in spring-time
vernici'flua ; varnish-exuding
veroni'cea ; Veronica-like
verruco'sa, um ; warty
verticilla'ris, verticilla'ta ; whorled
vesica'rium ; having bladders

EXPLANATION OF SPECIES-NAMES.

vesiculo′sum ; having little bladders
vesperti′lio; a bat
vesti′ta ; clothed
vexillif′era ; standard-bearing
villif′era ; wool-bearing
vilo′sa ; woolly
vimina′lis ; "osier"-like
vimin′ea ; osier-like
virga′ta ; twiggy
Virgin′icus; belonging to Virginia
vir′idis ; green
viscid′ula ; somewhat sticky
vis′cidum ; sticky
visco′sa ; glued
vitta′ta ; banded with a filet
volu′bilis, e ; twining
vomerifor′mis; ploughshare-shaped
vulga′ris, vulga′tum ; common
Zeylan′ica ; Ceylonese
zygophylloï′des ; Zygophyllum-like

INDEX TO THE ORDERS & GENERA,

AND EXPLANATION OF GENERIC NAMES.

	PAGE
Abu'tilon; Gr. for mulberry tree	30, 32
Aca'cia; Gr. *akazo*, to sharpen, hence thorny as some of the earlier known species. Wattle	60, 72
Acæ'na; Gr. *akaina*, a thorn	86
Acantha'ceæ; from Acan'thus, a prickle	10, 154
Achnoph'ora; Gr. *achne*, chaff; *phoros*, carrying	109, 116
Achyran'thes; Gr. *achuron*, chaff; *anthos*, a flower	53, 55
Acian'thus; Gr. *akis*, a point; *anthos*, flower	161, 165
Acrot'riche; Gr. *akros*, summit; *thrix (trichos)*, hair; tips of corolla-lobes bearded	148
Actino'tus; Gr. *aktinos*, of a ray; *otus*, ear; rayed involucrum	102
Adenan'thos; Gr. *aden*, a gland; and *anthos*, a flower	81
Adian'tum; Gr. *adiantos*, not wetted, supposed to keep dry when immersed. Maiden-hair	200, 202
Adria'na; after Adrien de Jussieu, the eminent botanist	38, 41
Æschynom'ene; Gr. *aischunomai*, to be modest; some species sensitive	59, 66
Agropy'ron; Gr. *agros*, a field; *puros*, wheat	189, 196
Agros'tis; Gr. name for grasses	188, 194
Ai'ra; Gr. for a weed in wheat, "darnel"	188, 195
Aïzo'on; Gr. *aei*, always; *zoos*, alive	86
Aj'uga; Gr. *a*, without; *zygos*, a yoke; calyx regular	149, 150
Alisma'ceæ; from Alis'ma, Water Plaintain	11, 173
Alopecu'rus; Gr. fox-tail (grass)	188, 193
Alternan'thera; stamens alternating with (staminodia)	53, 55
Alys'sum; Gr. *a*, without; *lyssa*, rage; in reference to the supposed properties of allaying anger	15, 17
Alyx'ia; Gr. *alukos*, salty; grows by the sea	142
Amaranta'ceæ; from Amaran'tus, not decaying	6, 53
Amaryllid'eæ; from Amaryl'lis, the daffodil	10, 168
Amman'nia; after a botanical professor at St. Petersburg	88
Amper'ea; after Ampere, the distinguished mathematician	38, 40
Amphipo'gon; Gr. *amphi*, around; *pogon*, a beard	188, 195
Andropo'gon; Gr. *andros*, man's; *pogon*, a beard	187, 191
Angian'thus; Gr. *anggeion*, a cup, and a flower (actually a pappus)	112, 128
Anthisti'ria; Gr. for a species of grass	187, 192
Anthob'olus; Gr., garlanded with flowers	99
Anthocer'cis; Gr. *anthos*, a flower; *kerkis*, a ray	144, 146
Anthot'roche; Gr. *anthos*; and *trochos*, a wheel	144
Ao'tus; *a*, without; *otos*, of an ear, *i.e.* bracteole	58, 62

INDEX TO THE ORDERS AND GENERA.

	PAGE
Apheli'a; Gr. *apheleia*, simplicity...	177, 178
A'pium; Celtic, *apon*, water. Celery-plant	102, 105
Apocyn'eæ; from Apo'cynum, Dog's-bane	9, 142
Aristi'da; Lat. *arista*, the beard of an ear of corn	188, 194
Arthropo'dium; Gr. *arthron*, a joint; *pous podos*, a stalk	169, 171
Arun'do; Lat. a reed	189, 196
Asclepiad'eæ; from Ascle'pias, a Greek physician	7, 141
Asper'ula; Lat. somewhat rough, hispid...	107, 108
Aspi'dium; Gr. the indusium in the form of a shield (*aspidos*)	201, 202
Asple'nium; Gr. without spleen, in allusion to supposed medicinal properties. Spleenwort	201, 202
As'ter; Gr. *astron*, a star—Starwort	109, 113
Astreb'la; Gr. (awn) not twisted	189, 196
Astrolo'ma; Gr. *astron*, a star; *loma*, a fringe	... 147
Atala'ya; an Indian name	... 27
Athrix'ia; without hairs (on the receptacle)	111, 122
A'triplex; Lat. from the Greek *atraphaxis*, the herb orach	45, 46
Avicen'nia; after Avicennes, a Persian physician	155, 156
Azol'la; meaning unknown...	... 199
Babbag'ia; after a South Australian explorer	46, 52
Baeck'ea; after a Swedish physician	89, 91
Bank'sia; after Sir Joseph Banks, botanist and companion of Capt. Cook	81, 85
Barbare'a; anciently called Herb St. Barbara	15, 16
Bas'sia; after F. Bassi, botanical curator at Bologna	46, 50
Baue'ra; after F. Bauer, botanical artist to Flinders' expedition	... 85
Bauhin'ia; after two celebrated botanists of the 16th century	60, 71
Ber'gia; after a Swedish botanical author (1757-80)	... 21
Ber'tya; after Count L. de Lambertye	38, 40
Beyer'ia; after a Dutch botanist	38, 40
Bi'dens; Lat. two-toothed (achenes)	108, 113
Bignonia'ceæ; from Bigno'nia, a personal name	10, 154
Billardie'ra; after Labillardiere, author of Nov. Holl. Plantarum	... 20
Blyx'a; *bluxo*, gushing (forth in water)	... 160
Boerhaav'ia; after a friend and patron of Linnæus	... 55
Boragin'eæ; from Bora'go, the Borage-plant	9, 139
Boro'nia; after F. Borone, an Italian botanical collector	... 22
Bossiæ'a; after Bossieu Lamartiniere, a companion of La Perouse	58, 65
Botrych'ium; fertile frond resembling a *bunch of grapes*	... 200
Brachych'iton; Gr. *brachys*, short; *chiton*, an inner coat	... 3, 35
Brachyc'ome; Gr. short; and *koma*, a head of hair	119, 117
Brachylo'ma; Gr. short; and *loma*, a fringe	... 146
Brachyse'ma; short; and *sema*, a shield (or standard)...	57, 61
Brewer'ia; after a friend of Dillenius	137, 138
Bro'mus; Lat. for wild-oat...	189, 196
Bruno'nia; Latanized name of Robert Brown, the eminent botanist	... 132
Buechne'ra; after a botanical author	... 152
Bul'bine; having *bulb-like* tubers	169, 171
Burchar'dia; personal name	169, 170

	PAGE
Bursa'ria ; Lat. *bursa*, a pouch, from the shape of the fruit	20
Burto'nia ; personal name	58, 61
Cæ'sia ; after F. Cæsius (1703)	169, 171
Cæsalpinie'æ ; from Cæsalpin'ia, after the Italian botanist Caesalpinuis (1583)	57
Caki'le ; Arabic for the "sea-rocket"	15, 17
Calade'nia; Gr. *kalos*, beautiful ; *aden*, a gland	161, 166
Calecta'sia ; Gr. beautiful extension	169, 172
Calda'sia ; after a Chilian botanist	102, 104
Calliste'mon ; Gr. *kallistos*, most beautiful ; and *stemon*, a stamen	89, 92
Callit'riche ; *kallistos* ; and *thrix trichos*, a hair	100, 102
Callit'ris ; leaves *most beautifully* arranged in *threes*	159
Caloceph'alus; *kalos*, beautiful ; and *cephale*, a head	112, 129
Calochi'lus ; Gr. *kalos* ; and *cheilos*, a lip or labellum	161, 163
Calog'yne ; Gr. *kalos* ; and *gyne*, a pistil or stigma	133, 136
Calostem'ma ; Gr. *kalos* ; and *stemma*, a crown	168
Calost'rophus ; Gr. a rope-twister	178
Calo'tis ; Gr. *kalos* ; and *ous otos*, an ear (pappus)	109, 116
Cal'ycothrix ; Gr. *calyx*-lobes extending in *hairs*	89, 90
Campanula'ceæ ; from Campanula—Bell-flower	8, 131
Candol'lea, ; after the distinguished Swiss botanist	130
Candollea'ceae ; from Candollea, after DeCandolle	8, 130
Can'thium ; from a Malabar name of a species	107
Capparid'eæ ; from Capparis. Caper-plant	3, 15
Cap'paris ; a latinized Arabic word	15
Caprifolia'ceæ ; from Caprifolium	8, 108
Capsel'la ; diminutive of *capsa*, a box	16, 17
Cardam'ine ; Gr. for a water-cress	15, 16
Ca'rex ; Lat. for a sedge	180, 184
Caris'sa ; derivation unknown	142
Caryophyl'leæ ; from D. Caryophyllus, clove-pink	6, 42
Cas'sia ; Lat. for an Arabian spice	60, 71
Cassin'ia ; after H. Cassini, a celebeated French botanist	111, 123
Casuar'ina ; from the resemblance of the branchlets to the feathers of the Cassowary (Casuarius)	56
Casuarin'eæ ; from Casuarina. Native Oaks	6, 56
Cassy'tha ; Gr. name of the dodder-plant	14
Catosper'ma ; Gr. *kata*, pendulous ; and *sperma*, a seed	132
Caus'tis ; Gr. scorched (in appearance)	180, 185
Centip'eda ; Gr. hundred feet, in allusion to creeping habit	110, 121
Centrol'epis ; Gr. *kentron*, a spur ; and *lepis*, a scale	177, 178
Centun'culus ; Lat. for small weed. Pimpernel	137
Cephalip'terum ; Gr. *cephale*, the head ; and *pteron*, wing	112, 129
Ceratog'yne ; Gr. *keras, keratos*, a horn ; and *gyne*, a pistil	110, 121
Ceratophyl'leæ ; from Ceratophyllum, Hornwort	3, 14
Ceratophyl'lum ; Gr. *keratos* ; and *phyllon*, a leaf	14
Chamæscil'la ; Gr. *chamai*, dwarf ; and *scilla*, a squill	169, 172
Cheilan'thes ; *cheilos*, a lip ; and *anthos*, a flower ; in allusion to the indusium	201, 203

INDEX TO THE ORDERS AND GENERA. 289

PAGE

Cheiran'thera; Gr. *cheir*, the hand; and *anthera*, an anther;
  in allusion to the one-sided stamens 20
Chenopodia'ceæ; from Chenopodium. Goose-foot 6, 45
Chenopo'dium; from the shape of the leaves of some species ... 45, 49
Chlo'ris; Gr. pale-green, from colour of herbage 188, 193
Chore'trum; meaning unknown 99
Chorizan'dra; Gr. stamens apart by themselves... ... ... 180, 186
Chthonoceph'alus; Gr. *chthone, chthonos*, the ground; and
  cephale, the head; flower-heads close to the ground ... 112, 130
Cla'dium; Gr. *klados*, a branch; inflorescence branched ... 180, 185
Clayto'nia; after a botanical collector 41
Clem'atis; Gr. *klema*, a twig easily broken 13
Cleo'me; Gr. *kleio*, to shut; of uncertain application 15
Cleroden'drum; Gr. *kleros*, a lot or chance, and *dendron*, tree;
  of uncertain application 155, 156
Clian'thus; Gr. glory-flower 59, 67
Codoncar'pus; Gr. *kodon kodonos*, a bell; and *carpos*, fruit 29
Colde'nia; after an American botanist (1742) 139
Coloban'thus; Gr. *kolobos*, curtailed, and *anthos*; petals absent ... 42
Comesper'ma; Gr. *kome*, a head of hair; and *sperma*, a seed ... 21
Commeli'na; after the brothers Commelins, Dutch botanists... ... 173
Commelin'eæ; from Commelina 11, 173
Commerço'nia; after the botanist to Bougainville's expedition 35, 36
Compos'itæ; from the composite or compound flowers ... 8, 108
Conif'eræ; from coniferous, cone-bearing 10, 159
Conosper'mum; Gr. *konos*, a cone; and *sperma*, a seed ... 81, 82
Convolvula'ceæ; from Convolvulus 8, 137
Convol'vulus; lat. *convolvo*, to entwine, in allusion to habit ... 137, 138
Copros'ma; Gr. *kopros*, dung, and *osme*, smell 107
Cor'chorus; the Greek name of a culinary vegetable 34
Corre'a; after Correa de Serra, a Portugese botanist 22
Corynothe'ca; Gr. *korune*, a club; and *theke*, a case (fruit) ... 169, 172
Corysan'thes; Gr. *korus*, a helmet; and *anthos*, a flower ... 161, 164
Cot'ula; Gr. *kotule*, a small cup, form of involucre 110, 120
Crantz'ia; after a botanical author (1762-68) 102, 104
Craspe'dia; Gr. leaves having a fringe (*kraspedon*) of hairs ... 112, 129
Crassula'ceae; from *crassulus*, somewhat thick, in allusion to
  the succulent leaves 7, 85
Cres'sa; from *Cressus*, appertaining to Crete 137, 138
Cri'num; Gr. *krinon*, a lily 168
Crotala'ria; Gr. *krotalon*, a rattle; seeds rattling in the ripe
  pod 59, 66
Crucif'eræ; cross-bearing, petals cruciate 3, 15
Cryptan'dra; Gr. *kruptos*, hidden; *andre*, stamens 96
Cryptosty'lis; Gr. *kruptos*; and *stulos*, the column 161, 163
Cu'cumis; Lat. for a cucumber 105
Cucurbita'ceæ; from Cucur'bita, a gourd... ... ... 8, 105
Cus'cuta; an Arabic word for "dodder" 137, 139
Cycad'eæ; from Cy'cas 10, 160

290 FLORA OF SOUTH AUSTRALIA.

	PAGE
Cymbi'dium; boat-shaped (labellum)	160, 162
Cymbono'tus; Gr. hollow-backed; form of achenes	109, 118
Cymodo'cea; after the name of a sea-nymph	175, 176
Cynan'chum; dog-strangler, in allusion to poisonous properties	141
Cyn'odon; Gr. *kuneos*, dog-like, and *odous*, tooth	188, 195
Cynoglos'sum; Gr. dog-like tongue. "Hound's-tongue"	139, 141
Cypera'ceæ; from Cyperus	11, 179
Cype'rus; Gr. *kupeiros*, a rush	171, 181
Cyrtosty'lis; Gr. *kurtos*, curved; and column	161, 166
Da'mia; an Arabic word	141
Damaso'nium; Gr. *damasis*, subduing; in illusion to supposed medical qualities	173
Dampie'ra; after Dampier, an early Australian navigator	132
Dantho'nia; after M. Danthoine, a French botanist	189, 196
Darwin'ia; after the celebrated Dr. Darwin	89, 90
Datu'ra; Arabic name for a "thorn-apple"	144
Dau'cus; Gr. *daukon*, a wild carrot	103, 105
Davie'sia; after a Welsh botanist	58, 61
Dentel'la; diminutive of *dens, dentis*, a tooth	106
Dianel'la; dim. of Diana, the goddess of hunting	169, 171
Dichelach'ne; Gr. *dis*, double; *cheilos*, a lip; and *achne*, chaff—glumes, two-lobed	188, 194
Dichon'dra; Gr. *di*, two, and *chondros*, a grain	137, 138
Dickso'nia; after a Scotch cryptogamic botanist	200
Dicrasty'lis; Gr. *dicranos*, 2-headed, and style	155
Didis'cus; Gr. two-discs, alluding to the fruitlets	102, 104
Didymothe'ca; Gr. *didymos*, double, and *theke*, seed-vessel	29
Dillenia'ceæ; from Dillenia, after Dillenius, botanical professor at Oxford	3, 14
Dillwyn'ia; after an English botanist	58, 62
Dimorphoc'oma; Gr. *dis*, two, *morphe*, form, and *kome*, pappus	109, 115
Diplach'ne; Gr. *diploos*, double, and *achne*, chaff or glume	189, 197
Diplopel'tis; Gr. *diploos*, and *pelte*, a small shield	27
Dipo'dium; Gr. *dipodes*, two feet; in allusion to the two-stalked pollen masses	160, 162
Di'stichlis; Gr. *distichos*, (leaves) in two rows	189, 197
Diu'ris; Gr. *dis*, two, and *oura*, a tail; referring to the elongate lateral sepals	161, 163
Dodonæ'a; after a botanical writer (1578-1585)	27
Dro'sera; Gr. *droseros*, dewy	19
Drosera'ceæ; from Drosera, Sun-dew	4, 19
Dryma'ria; inhabiting, *drymos*, a forest	42
Duboi'sia; personal name	144, 146
Dyspha'nia; Gr. badly visible, *i.e.* the flowers	45, 48
Echinopo'gon; Gr. *echinos*, (prickly as) the hedge-hog, and *pogon*, the beard	188, 192
Echinosper'mum; Gr. *echinos*, and *sperma*, a seed	139, 140
Ehrhar'ta; after a Swiss botanist	187, 192
Elachan'thus; Gr. *elachus*, short, and *anthos*, a flower	110, 121
Elat'ine; Gr. *elate*, a fir	21

INDEX TO THE ORDERS AND GENERA. 291

	PAGE
Elatin'eæ; from Elatine	4, 21
Eleusi'ne; Elusis, another name for Ceres	189, 197
Elytroph'orus; Gr. *elutron*, a sheath, and *phoros*, bearing; referring to the outer glume	188, 196
Encephalar'tos; Gr. in the head bread; top being edible	160
Enchylæ'na; Gr. *enchylos*, succulent, and *chlaena*, a cloak	46, 49
Epacrid'eæ; from Epacris, Gr. *epi*, upon, and *akros*, the top	9, 149
E'pacris; in allusion to the mountain habit	146, 148
Epal'tes; meaning unknown	110, 121
Epilo'bium; Gr. calyx-lobes *upon* the *pod*	88
Eragros'tis; Gr. *eros*, love, and *agrostis*. Love-grass	184, 197
Erechthi'tes; a Greek name for a species of Senecio	109, 119
Eremoph'ila; Gr. *eremos*, a desert, and *philos*, a lover	156
Eriach'ne; Gr. *erion*, woolly, and *achne*, chaff	188, 195
Erian'thus; Gr. *erion*, woolly, and *anthos*, a flower	187, 191
Eriochi'lus; Gr. *erion*, woolly, and *cheilos*, a lip	161, 166
Eriochlam'ys; Gr. *erion*, and *chlamys*, a cloak	112, 129
Erioch'loa; Gr. *erion*, and *chloa*, a blade of grass	187, 192
Erioste'mon; Gr. *erion*, and *stemon*, a stamen	22, 24
Eritrich'ium; Gr. *erion*, and *thrix trichos*, a hair	139, 141
Erodiophyl'lum; Gr. *Erodium*, and *phyllon*, a leaf	109, 118
Ero'dium; Gr. *erodios*, a heron. "Heron's bill"	26
Eryn'gium; Gr. *ereugo*, to eructate, in allusion to supposed properties	102, 104
Ery'simum; Gr. *erusis*, a blistering. Blister-plant	15, 16
Erythræ'a; Gr. *eruthros*, red (flowers)	142
Eryth'rina; Gr. *eruthraino*, to dye red	60, 70
Ethuliop'sis; Gr. Ethulia (an allied genus) -like	110, 121
Eucalyp'tus; Gr. *eu*, well; *kalupto*, to cover as with a lid	89, 93
Euphor'bia; after Euphorbos, a Numidian physician	37, 38
Euphorbia'ceæ; from Euphorbia	5, 37
Euphra'sia; altered from Euphrosyne, one of the three graces, expressing gladness	152
Eutax'ia; Gr. well-behaved, modesty	58, 62
Euxo'lus; Gr. *euxulos*, juicy, well-flavoured	53
Exocar'pos; Gr. *exo*, without, and *karpos*, fruit	99, 100
Evol'vulus; not twining in contradistinction with Convolvulus	137, 138
Festu'ca; Celtic, *fest*, pasture. Fescue-grass	189, 197
Ficoïd'eæ; plants agreeing with the *fig-like* marigold	7, 86
Fi'cus; Lat. a Fig-tree	56
Fi'lices; Lat. Ferns	12, 200
Fimbristy'lis; Lat. having a fringed-style	180, 183
Flave'ria; the Chilian species yielding a *yellow* dye	108, 113
Fluvia'les; plants belonging to rivers	11, 174
Franke'nia; after Prof. Frankenius, of Upsal (1638-1661)	20
Frankenia'ceae; from Frankenia	4, 20
Fuire'na; personal name	
Fu'sanus; ancient name of the *Spindle*-tree	99
Galac'tia; Gr. *galaktos*, of milk; some species with a milky juice	60, 70

	PAGE
Ga'lium ; Gr., one species used for curdling, *gala*, milk	107, 108
Gastrolo'bium ; Gr. *gastros*, of the belly (inflated), and *lobos*, a pod	58, 63
Geijer'a ; after a botanical author	22, 24
Gentia'na ; after Gentius, a king of Illyria	142
Gentian'eæ ; from Gentiana	9, 142
Geococ'cus ; Gr. earth-berry	15, 17
Gerania'ceæ ; from Geranium	5, 26
Gera'nium ; Gr. *geranos*, a crane. Crane's-bill"	26
Ge'um ; Gr. *geno*, to give a relish ; root aromatic	86
Gleiche'nia ; after a German cryptogamic botanist	200
Glosso'dia ; Gr. tongue-like (glands on base of labellum)	161, 166
Glossog'yne ; Gr. *glossa*, a tongue, and *gyne*, pistil	108, 113
Glossostig'ma ; Gr. *glossa*, and stigma	152
Glyc'ine ; Gr. *glukus*, sweet, as roots of some species	60, 70
Glycyrrhi'za ; Gr. *glukus*, and *rhiza*, a root. Liquorice	59, 66
Gnapha'lium ; Gr. *gnaphalon*, soft-down, clothing the plants	111, 123
Gnaphalo'des ; Gr. Gnaphalium-like	112, 129
Gnepho'sis ; Gr. in reference to the jagged pappus-cup of the type-species	112, 128
Gompholo'bium ; Gr. *gomphos*, a wedge-shaped nail, and *lobos*, a pod	57, 61
Gomphre'na ; altered from Gromphrena, an ancient name for a plant. Globe-amarant	53, 55
Goode'nia ; after Dr. Bishop Goodenough, a botanical author	133, 135
Goodenia'ceæ ; from Goodenia	8, 132
Good'ia ; after a botanical collector	59, 66
Gossyp'ium ; Arabic, in allusion to the cottony seeds	30, 34
Gramin'eæ ; from *gramineus*, grassy. Grasses	12, 187
Grammi'tis ; Gr. *gramma*, a writing, referring to the arrangement of the sori	201, 202
Grati'ola ; *gratiosus*, in great esteem	152
Grevil'lea ; after a great promoter of natural history	81, 82
Gun'nia ; after a Tasmanian botanist	86
Gyroste'mon ; Gr. stamens arranged in a circle	29
Ha'kea ; after Baron Hake, a botanical patron	81, 83
Halga'nia ; after a French admiral	139, 140
Haloph'ila ; Gr. *halos*, of the sea, and *philos*, a lover	160
Halorag'eæ ; from Haloragis	7, 100
Halora'gis ; *halos*, and *rax ragis*, a berry	101
Hannafor'dia ; after an Australian naturalist	35, 36
Heleoch'aris ; Gr. *heleios*, marshy, *chairo*, to delight in	180, 183
Helichry'sum ; partly in allusion to the brilliant yellow (*chrysos*) colour, applicable to such species as *H. lucidum*	111, 126
Heliotro'pium ; Gr. *helios*, the sun, and *trope*, a turning round	139
Helip'terum ; by ellipsis from Helichrysum, and *pteros*, a wing, in allusion to the feathery pappus	111, 124
Hemarth'ria ; Gr. half-jointed, referring to the rhachis	187, 191
Herman'nia ; after Prof. Hermann, of Leyden, died 1695	35
Hernia'ria ; *hernia*, a rupture, which it is imagined to cure. "Rupture-wort"	44

INDEX TO THE ORDERS AND GENERA. 293

	PAGE
Heteroden'dron; Gr. *heteros*, variable; *dendron*, a tree	27
Hibber'tia; after a botanical patron	14
Hibis'cus; Gr. for a mallow-plant	30, 33
Ho'vea; after a Polish botanist	59, 66
Howit'tia; after Dr. G. Howitt, of the Burke and Wills' relief expedition	30, 32
Hu'mea; after Sir A. Hume, a botanical patron	110, 121
Hyalol'epis; Gr. *hyaleos*, transparent; *lepis*, a scale	111, 128
Hyban'thus; Gr. *hubos*, a hump; *anthos*, a flower	18, 19
Hydril'la; dim. of *hydra*, a water-serpent	160
Hydrocharid'eæ; from Hydrocharis, delighting in water	10, 160
Hydrocot'yle; Gr. *hudor*, water; *cotule*, a small cup	102
Hymenan'thera; Gr. *hymen*, a membrane; *anthera*, an anther	18, 19
Hypericin'eæ; from Hypericum	4, 22
Hyper'icum; Gr. *huper*, above; *eicon*, an image; superior part of the flower resembles a figure	22
Hypox'is; Gr. capsule pointed (*oxus*) below (*hupo*)	168
Illecebra'ceæ; from Illecebrum, a name given by Pliny	6, 44
Impera'ta; after Ferranti Imperati	187, 191
Indigof'era; Lat. *indigo*, a blue dye, and *fero*, to bear	59, 67
Ipomœ'a; Gr. *ipos*, bindweed, and *omoios*, like	137
Irid'eæ; from Iris, the rainbow. "Flag"	10, 167
Isoëtop'sis; Gr. Isoëtes-looking, as regards foliage	110, 118
Isol'epis; *isos*, equal; *lepis*, a scale	180, 184
Isopo'gon; Gr. *isos*, equal, *pogon*, a beard	81
Isot'oma; Gr. from its equally cut corolla	131, 132
Isot'ropis; Gr. *isos*, equal, *tropis*, a keel	57, 61
Ixiolæ'na; Gr. *ixos*, birdlime, *chlaina*, a cloak	111, 122
Ixo'dia; Gr. sticky *like birdlime*	110, 121
Jasmin'eæ; from Jasminum	9, 143
Jasmin'um; from the Arabic word *jasmin*	143
Josephi'nia; after the Empress Josephine	154
Junca'ceæ; from Juncus. Rushes	11, 173
Jun'cus; Lat. *jungo*, to join; made into ropes	173
Jussieu'a; after the uncle of the celebrated French systematic botanist, A. L. Jussieu	88
Justic'ia; after an eminent Scotch agriculturist	154
Kenne'dya; after an English nurseryman	60, 70
Ko'chia; after Dr. W. D. J. Koch, a German botanist	46, 49
Kun'zea; after a botanical professor at Leipsic	89, 92
Kyllin'gia; after a Danish botanist, died 1696	179, 180
Labia'tæ; from *labium* a lip, corolla usually 2-lipped	9, 148
Lagenoph'ora; Gr. *lagena*, a flask, *phoros*, carrying	109, 117
Lasiopet'alum; Gr. *lasios*, woolly, *petalon*, a petal	35, 36
Laura'ceæ; from Laurus, a laurel	3, 14
Lavate'ra; after a naturalist of Zurich	30, 33
Laxman'nia; after a Siberian traveller	169, 172
Leewenhoe'kia; after the Dutch philosopher and micrographist (1683)	130
Legumino'sæ; from *legume*, a pod	6, 57

	PAGE
Lem'na; Gr. *Lemnos*, an island in the Ægean Sea	177
Lemna'ceæ; from Lemna; "Duckweed"	11, 177
Lentibularin'eæ; from Lentibularia	9, 151
Lepid'ium; Gr. *lepis lepidos*, a scale; in allusion to the scale-like fruit	16, 18
Lepidob'olus; Gr. *lepidos* and *bolos*, anything thrown	178
Lepidosper'ma; Gr. *lepidos*, and *sperma* a seed	180, 186
Lepidos'pora; Gr. *lepidos*, and *spora*, seed	179, 180
Lepilæ'na; Gr. *lepis*, and *chlaina*, a coat (bract)	175, 176
Leptocar'pus; Gr. *leptos*, thin, *carpos*, fruit	178
Leptome'ria; Gr. having slender parts (branches)	99
Leptosper'mum; Gr. *leptos*, slender, *sperma*, a seed	89, 91
Leptorrhyn'chos; Gr. *leptos*, and *rhunchos*, a beak	111, 124
Leptu'rus; Gr. *leptos*, and *oura*, a tail; in allusion to the pointed rhachis	187, 192
Lepyro'dia; Gr. having *husk-like* bracts	177, 178
Leschenaul'tia; after Leschenault de la Tour, one of the botanists to Baudin's Expedition	132
Lespede'za; after Lespedez, a botanical patron	60, 69
Leucopo'gon; Gr. corolla-lobes *bearded with white* hairs	147
Lhotz'kya; after a German botanist	89, 90
Lilia'ceæ; from Lilium, a lily	11, 169
Limnan'themum; Gr. *limne*, a marsh, *anthemos*, flowery	142
Limosel'la; diminutive of *limosus*, muddy	152
Lind'sæa; after an eminent English botanical professor	200
Lin'eæ; from Linum, flax-plant	4, 25
Li'num; Lat. a thread	25
Lipocar'pha; having shining bracts	180, 184
Lissan'the; Gr. having smooth (not bearded) flowers	147
Livisto'na; after Patrick Murray of Livistone	174
Lobe'lia; after L'Obel, a botanical author (1538-1616)	131
Loga'nia; after a botanical writer (1747)	143, 144
Logania'ceæ; from Logania	9, 143
Loma'ria; Gr. *loma*, a fringe or edge; from the position of the indusium	201, 202
Lorantha'ceæ; from Loranthus. Mistletoe	8, 105
Loran'thus; *lorum*, a thong, *anthos*, a flower	105, 106
Lo'tus; Gr. for a clover- or trefoil-plant	60, 69
Loudo'nia; after J. C. Loudon, a botanical author	100
Lu'zula; *gramen luzulæ*, glow-worm grass of ancient botanists	173
Lyc'ium; original species, native of Lycia	144
Lycopodia'ceæ; from Lycopodium. Club-moss	12, 199
Lycopo'dium; Gr. *lukos*, a wolf, *pous podos*, a foot	199
Lyco'pus; Gr. *lukos*, a wolf, and *pous*, a foot	149, 150
Lyperan'thus; Gr. *luperos*, troublesome, and *anthos*; colour of flowers difficult to preserve	161, 166
Lythra'ceæ; from Lythrum. Loose-strife	7, 88
Ly'thrum; Gr. gore, referring to the colour of the flowers	88
Macgrego'ria; after a Victorian senator	29
Malva'ceæ; from Mal'va, a mallow	5, 30

INDEX TO THE ORDERS AND GENERA. 295

	PAGE
Malvas'trum ; altered from Malva...	30, 33
Marian'thus ; Marie-flower ...	... 20
Marsde'nia ; after the author of the "History of Sumatra" ...	... 141
Marsi'lea ; after Count L. F. Marsigli ...	... 199
Ma'zus ; Gr. *mazos*, a breast ; from the protuberance in the corolla-throat ...	... 152
Melaleu'ca ; Gr. black-white, from the colour of the trunk and branches of original species ...	89, 92
Melha'nia ; from Mt. Melhan in Arabia ...	... 35
Melia'ceæ ; from Me'lia, the Greek for the Ash ...	... 4, 25
Meloth'ria ; Gr. for a melon-like plant ...	... 105
Men'kea ; a personal name ...	15, 17
Men'tha ; Latin for the mint-plant ...	... 148
Mesembrian'themum ; Gr. mid-day flowering ...	... 86
Micran'theum ; Gr. *micros*, small ; *antheion*, a blossom ...	37, 39
Microco'rys ; Gr. *micros*, and *korus*, a helmet ...	149, 150
Micros'eris ; Gr. *micros*, and *seris*, lettuce ...	112, 130
Micro'tis ; Gr. *micros*, and *otis*, ear ...	161, 164
Millo'tia ; a personal name ...	110, 119
Mimo'seæ ; from Mimo'sa, a mimic ; leaves sensitive ...	... 57
Mim'ulus ; Gr. *mimo*, an ape ; from the appearance of the seeds. Monkey-flower...	... 152
Minu'ria ; meaning unknown ...	109, 115
Mirbe'lia ; after a distinguished physiological botanist ...	58, 61
Mitrasac'me ; Gr. *mitra*, a turban (capsule) ; *akme*, a point ...	... 143
Mollu'go ; Pliny's name for a herb ...	... 86
Momor'dica ; Lat. *mordeo*, to bite ; the seeds have the appearance of being bitten ...	... 105
Monotax'is ; Gr. *monos*, one ; *taxis*, a row ...	38, 40
Monot'oca ; Gr. *monos*, alone, *tokos*, birth ; 1-ovulate ...	... 148
Muehlenbec'kia ; after an Alsatian botanist ...	... 44
Myoporin'eæ ; from Myop'orum, Native Myrtle ...	10, 156
Myop'orum ; Gr. closed pores, in allusion to the leaf-glands ...	... 156
Myoso'tis ; Gr. *mus*, mouse ; *ous*, *otis*, an ear ; from shape of leaves ...	139, 141
Myosu'rus ; Gr. *mus*, and *oura*, a tail ...	... 13
Myriophyl'lum ; Gr. *murious*, numerous (divided), and leaf ...	100, 101
Myrta'ceæ ; from Myrtus, a myrtle ...	... 7, 89
Na'ias ; a water-nymph ...	175, 177
Nastur'tium ; *nasus*, the nose, *tortus*, tormented ...	15, 16
Nematophyl'lum ; Gr. *nema*, a thread, and *phyllon*, a leaf ...	59, 66
Neptu'nia ; after the mythological god of the sea ...	60, 72
Neurach'ne ; Gr. *neuron*, a nerve ; *achne*, chaff (glumes) ...	187, 191
Newcast'lia ; after the Duke of Newcastle ...	... 155
Nicotia'na ; after J. Nicot, the introducer to Europe of the tobacco-plant ...	... 144
Nitra'ria ; Gr. *nitron*, salt, from the nature of habitat...	25, 26
Notone'rium ; Gr. *nothos*, spurious ; Nerium, the oleander ...	... 142
Nyctagin'eæ ; from Nyctago, Night-bell ...	... 6, 55
Olacin'eæ ; from Olax ...	... 7, 98

	PAGE
O'lax ; Gr. *aulax*, a furrow ; but not applicable	98
Oldenlan'dia ; after a botanical collector (1695)	106
Onagre'æ ; from Onagra, an old name for Œnothera	7, 88
Opercula'ria ; in reference to the calyx-lid	107
Ophioglos'sum ; Gr. *ophis*, a serpent ; *glossus*, a tongue	200
Orchid'eæ ; from Or'chis	10, 160
Oroban'che ; Gr. *orobos*, a vetch, *ancho*, to strangle	151
Oroban'cheæ ; from Orobanche. Broom-rape	9, 151
Orthoc'eras ; Gr. lateral sepals like *straight horns*	161, 163
Osmun'da ; after a Celtic divinity	200
Otte'lia ; a personal name	160
Owe'nia ; a personal name	25
Ox'alis ; Lat. for wood-sorrel	26, 27
Pal'mæ ; from Palmus, a palm-tree	11, 174
Pan'icum ; Lat. for millet, used as bread *(panis)*	187, 190
Papa'ver ; Lat. *papa* ; thick milk, contained in the stem	14
Papavera'ceæ ; from Papaver. Poppy	3, 14
Papiliona'ceæ ; from *papilio*, *onis*, a butterfly	57
Pappaph'orum ; Gr. pappus or beard carrying	187, 192
Parieta'ria ; because the type species grows on walls	56
Paterso'nia ; after an early New South Wales botanist	167, 168
Pedalin'eæ ; from Pedalium	10, 154
Pelargon'ium ; Gr. *pelargos*, a stork. Stork's-bill	26
Pennise'tum ; Lat. *penna*, a feather, *seta*, a bristle	187, 190
Pentapo'gon ; Gr. having five beards or awns	188, 195
Peplid'ium ; Pep'lis, Gr. name for Purslane, and *idea*, like	152, 153
Pero'tis ; Gr. *much eared* or awned	187, 192
Persoon'ia ; after the author of a "Synopsis Plantarum"	81, 82
Petalosty'lis ; Gr. having a petaloid style	60, 71
Petroph'ila ; Gr. rock-loving	81
Phyllan'thus ; in allusion to the flowers growing on the edge of leaf-like branches in some species	38, 39
Phyllo'ta ; Gr. having ear-shaped leaves	58, 62
Phytolacce'æ ; from Phytolac'ca	5, 29
Pilula'ria ; bearing little pills. Pillwort	199, 200
Pimele'a ; Gr. *pimela*, fat ; in allusion to the oily seed	79
Pittospor'eæ ; from Pittosporum	4, 20
Pittos'porum ; Gr. *pitte*, resin, and *sporos*, seed	20
Plagian'thus ; Gr. *plagios*, oblique, and *anthos*, a flower	30
Plantagin'eæ ; from Plantago	9, 143
Planta'go ; plantain	143
Platylo'bium ; Gr. *platys*, broad, and *lobos*, a pod	58, 65
Plectran'thus ; Gr. *plectron*, a spur, and *anthos*	149, 150
Pluche'a ; after M. Pluche	111, 122
Plumbagin'eæ ; from Plumbago, lead-wort	6, 55
Plumba'go ; from *plumbum*, lead	55
Po'a ; Gr. name of a herb	189, 198
Podoc'oma ; Gr. having a *stalked pappus*	109, 115
Podol'epis ; Gr. having the *scales* (or inner phyllaries) *on stalks*	111, 123
Podosper'ma ; Gr. stalk-seeded	111, 122

INDEX TO THE ORDERS AND GENERA.

	PAGE
Pollich'ia ; after a German botanical author	139, 140
Polycalym'ina ; Gr. having many coverings	111, 127
Polycarpæ'a ; only indicating its affinity to	42
Polycar'pon ; Gr. many-fruited. All-seed	42
Polycne'mon ; Gr. having many-kneed or jointed stems	53
Polyg'ala ; Gr. much-milk, from cows feeding on it	21
Polyga'leæ ; from Polygala	4, 21
Polygona'ceæ ; from Polygonum. Knot-grass	6, 44
Polyg'onum ; Gr. *polus*, many, *gonu*, knee or joint	44
Polyme'ria ; Gr. *polu-meres*, consisting of many parts	137, 138
Polypo'dium ; Gr. *polu-podes*, the many-footed	201, 202
Polypompho'lyx ; Gr. *polus*, many ; *pompho'lux*, a bubble	151
Pomader'ris ; Gr. for a cloak of undressed skin	96
Po'max ; Gr. *poma*, a lid (to the capsule)	107
Poran'thera ; having anthers opening by pores	37, 38
Portula'ca ; Lat. *porto*, to carry, *lac*, milk	7, 41
Portula'ceæ ; from Portulaca	6, 41
Posido'nia ; Gr. *Poseidon*, a god of the sea	175, 176
Potentil'la ; somewhat powerful, root astringent	96
Potamoge'ton ; Gr. *potamos*, a river ; *geiton*, near	175, 176
Prasophyl'lum ; Gr. *prason*, a leek, and *phyllon*, a leaf	161, 164
Primula'ceæ ; from Primula, a primrose	8, 137
Prostan'thera ; Gr. having appendaged anthers	149, 150
Protea'ceæ ; from Pro'tea (assuming different shapes)	7, 81
Prunel'la ; of doubtful derivation. Self-heal	149, 150
Pseudan'thus ; Gr. *pseudos*, false ; *anthos*, a flower	6, 37
Psora'lea ; Gr. *psoraleos*, scurfy	60, 64
Pterig'eron ; *pteron*, a wing ; *gero*, to bear	109, 111, 122
Pte'ris ; the bracken-fern	201, 202
Pterocau'lon ; Gr. *pteron*, a wing, *kaulos*, a stem	112, 129
Pterosty'lis ; Gr. *pteron*, and *stulos*, the column	161, 164
Ptilo'tus ; Gr. *ptilotos*, feathered	53
Ptychose'ma ; Gr. *ptuchos*, of a fold, *sema*, a standard	59, 66
Pultenæ'a ; after Dr. Pulteney, a botanical writer	58, 63
Quine'tia ; a personal name	110, 119
Ranuncula'ceæ ; from Ranunculus	3, 13
Ranun'culus ; diminutive of *rana*, a frog	13
Restia'ceæ ; from Restio ; *restis*, a cord	11, 177
Res'tio ; some species used for cordage	178
Rhamna'ceæ ; from Rhamnus. Buck-thorn	7, 96
Rhago'dia ; Gr. *ragodes*, bearing berries	45, 48
Rhizosper'mæ ; having rooting seeds or spores	12, 199
Rhyncho'sia ; Gr. having a beaked keel	60, 70
Ricinocar'pus ; Gr. Ricinus (castor-oil-plant) -fruited	38, 40
Roche'lia ; after a botanical author	139, 140
Rosa'ceæ ; from Rosa, the Rose	7, 85
Rota'la ; *rota*, a wheel ; leaves whorled	88
Rubia'ceæ ; from Ru'bia, the Madderwort	8, 106
Ru'bus ; Lat. a bramble	86
Ruel'lia ; after the author of De Natura Plantarum (1536)	154

U

	PAGE
Ru'mex ; Lat. a "Dock"	44
Rup'pia ; after H. B. Ruppius, a botanical author (1718)	175, 176
Ruta'ceæ ; from Ru'ta, the Rue-worts	4, 22
Rutid'osis ; Gr. *rutis*, *rutidos*, a wrinkle ; bracts wrinkled	110, 122
Sagi'na ; Lat. so called for its *nourishing* qualities	42
Salicor'nia ; *sal*, salt, and *cornu*, a horn. Marsh samphire	46, 52
Sal'sola ; *salsulus*, salted. Salt-wort	46, 52
Sambu'cus ; Lat. for elder-tree	108
Sam'olus ; Lat. for a marsh plant	137
Santala'ceæ ; from Santalum. Sandal-woods	7, 99
San'talum ; Arabic, *sandal*, useful. Sandal-wood	99
Sapinda'ceæ ; from Sapindus. Indian soap	5, 27
Sapona'ria ; yielding *sapo saponis*, soap	42
Sarcostem'ma ; Gr. *sarka sarkos*, fleshy ; *stemma*, a crown	141
Saxifra'geæ ; from Saxifra'ga	7, 85
Scæ'vola ; *scævus*, the left hand ; from shape of corolla	132, 134
Schedon'orus ; Gr. *schedon*, near ; *oros*, top ; awn from near the top of the flower-bract	183, 197
Schizæ'a ; Gr. *schizo*, to split ; fronds divided	200
Schœ'nus ; Gr. *schœnos*, a cord ; yielding cordage	179, 182
Scir'pus ; from *cirs*, a Celtic word for rushes	180, 183
Scleran'thus ; Gr. *skleros*, hard ; *anthos*, flower	44
Scrophularin'eæ ; from Scrophula'ria. Fig-wort	9, 152
Scutella'ria ; from *scutella*, a little saucer, the form of the calyx. Skull-cap	149, 150
Sebæ'a ; after Seba, a Dutch naturalist (1734-65)	142
Selaginel'la ; diminutive of Sela'go, a club-moss	199
Sellie'ra ; after Sellier, a Spanish artist	132
Sene'cio ; Lat. *senex*, an old man ; in allusion to the white pappus	109, 110, 119
Serin'gia ; after, N. C. Seringe, a Swiss botanist	35, 36
Sesba'nia ; "Sesban", the Arabic name of a species	59, 67
Seta'ria ; *seta*, a bristle ; referring to those on the rhachis	187, 189
Si'da ; Gr. for a mallow-like plant	30, 31
Siegesbeck'ia ; after the botanical curator at St. Petersburgh (1736)	108, 112
Sisym'brium ; Gr. *sisumbrion*, applied to a cress-plant	15, 16
Sisyrin'chium ; Gr. *sus*, a pig, *rhunchos*, snout	167, 168
Si'um ; Celtic, *siu*, water ; species semi-aquatic	102, 105
Skirroph'orus ; Gr. *skirros*, hard rind, *phoros*, bearing ; corolla hardened at the base	112, 128
Solana'ceæ ; from Solanum	9, 144
Sola'num ; Lat. for the Night-shade *(S. nigrum)*	144
Solenog'yne ; Gr. having a tubular pistil	109, 118
Spartotham'nus ; *Spartum*, "broom"; *thamnos*, a bush	155, 156
Spergula'ria ; allied to the genus Sper'gula	42
Spermaco'ce ; Gr. *sperma*, a seed, *akoke*, a point	107, 108
Sphærolo'bium ; Gr. *sphaira*, a globe ; *lobos*, a pod	58, 61
Spi'nifex ; Lat. spiny (leaves)	187, 191
Spiran'thes ; Gr. having spirally arranged flowers	161, 164

INDEX TO THE ORDERS AND GENERA. 299

	PAGE
Sporob'olus ; Gr. *sporos*, seed, *bolos*, sheading ; grains readily fall out	188, 195
Sprenge'lia ; after a distinguished Prussian botanist (1793)	146, 148
Spyrid'ium ; Gr. of the form of a round basket	97
Stackhou'sia ; after a British botanist	29
Stackhousi'eæ ; from Stackhousia	5, 28
Stella'ria ; in allusion to the star-like corolla	42
Stemo'dia ; *stemon*, a stamen, and *dis*, double ; anther-cells quite separate	152
Stenopet'alum ; Gr. *stenos*, narrow, *petalon*, a petal	15, 17
Sterculia'ceæ ; from Stercu'lia, after Sterculius	5, 35
Sti'pa ; Gr. *stupe*, tow ; in allusion to the feathery awns	188, 193
Stylid'ieæ ; from Stylidium, referring to the anthers	130
Stylid'ium ; connate with the style	130
Stylobas'ium ; Gr. referring to the basilary style	86
Styphe'lia ; Gr. *stuphelos*, hard, *i.e.* the leaves	146
Stuarti'na ; after McDougal Stuart, the explorer	110, 121
Suæ'da ; *suaed*, an Arabic word for a soda-plant	46, 52
Swainso'nia ; after a botanical patron of the 18th cent.	59, 67
Tec'oma ; a Mexican name for a species	154
Templeto'nia ; after an Irish naturalist	59, 65
Tephro'sia ; Gr. *tephros*, ash-coloured (leaves)	59, 67
Tetrago'nia ; Gr. having four-angled (fruits)	86, 87
Tetrathe'ca ; Gr. fourfold cases ; anthers 4-celled	37
Teu'crium ; after Teucer, a king of Troy	148
Thelym'itra ; Gr. for a woman's head-dress ; in allusion to the hooded column	161, 162
Thoma'sia ; after a Swiss botanist	35, 36
Threlkel'dia ; after Dr. Caleb Threlkeld	46, 49
Thryptome'ne ; Gr. *thrupto*, to break ; *mene*, a crescent	89, 90
Thyme'leæ ; from Thymelea. Spurge-laurels	6, 79
Thysano'tus ; Gr. *thusanos*, a fringe ; *ous*, *otis* an ear	169, 172
Tilia'ceæ ; from Til'ia, a Lime-tree	5, 34
Tillæ'a ; after M. A. Telli, a botanical professor at Pisa (1653-1740)	85
Toxan'thus ; Gr. *toxos*, a bow, and *anthos*, a flower	110, 118
Trachyme'ne ; Gr. *trachus*, rough ; *mene*, a crescent (fruitlet)	102, 104
Tra'gus ; bearded like a *goat*	187, 192
Tre'ma ; a hole, alluding to the pitted endocarp	post.
Treman'dreæ ; from Treman'dra	5, 37
Trian'thema ; Gr. having three flowers together	86
Tri'bulus ; Gr. three-pointed	25
Tricory'ne ; Gr. with triple club (-shaped fruit)	169, 172
Triglo'chin ; Gr. three-pointed (fruit)	174
Trigonel'la ; diminutive of *trigonus*, three-cornered	60, 69
Trio'dia ; Gr. having *three-toothed* glumes	189, 197
Tri'raphis ; Gr. with three needles ; three-awned glumes	188, 196
Tri'thuria ; Gr. *treis*, three ; *thurion*, a little door	177, 178
Triumfet'ta ; after an Italian botanist, died 1707	34
Trymal'ium ; Gr. *trūmalia*, eye of a needle	97

Ty'pha ; Lat. for the bull-rush	174
Typha'ceæ; from Typha	11, 174
Umbellif'eræ ; bearing umbels	7, 102
Urti'ca ; Lat. a nettle	56
Urtica'ceæ ; from Urtica	6, 56
Utricula'ria ; Lat. having little bladders	151
Vallisne'ria ; after A. Vallisneri, an Italian botanist	160
Velle'ya ; after Major Velley, a cryptogamic botanist	132, 133
Ventila'go ; Lat. *ventilo*, to blow, and *ayo*, to drive away ; the fruits are winged.	96
Verbe'na ; Lat. for the Vervain-plant	155
Verbena'ceæ ; from Verbena	10, 155
Veron'ica ; perhaps a feminine proname	152, 154
Verticor'dia ; Lat. that turneth the heart	89, 90
Vig'na ; after Vign, a commentator on Theophrastus	60, 70
Vimina'ria ; Lat. twiggy ; osier-like branchlets	58, 61
Vi'ola ; Lat. a violet	18
Viola'ceæ ; from Viola	4, 18
Vis'cum ; Lat. sticky (berries). Mistletoe	105, 106
Vittadi'nia ; after C. Vittadini, an Italian botanist	109, 115
Wahlenber'gia ; after the author of Fl. Lapponica (1812)	131, 132
Wait'zia ; personal name	111, 126
Walthe'ria ; after a German botanist (1735)	35
Wede'lia ; after a botanical professor at Jena (1625-1721)	108, 112
Westrin'gia ; after a Swedish physician	149, 150
Wilso'nia ; after the author of a Synopsis of British Plants	137, 138
Wol'ffia ; personal name	177
Wurmbe'a ; after F. von Wurmb, a Batavian botanist	169, 170
Xanthorrhœ'a ; Gr. exuding yellow (resin)	170, 172
Xantho'sia ; Gr. *xanthos*, yellow ; the colour of the hairs	102, 104
Xero'tes ; Gr. dryness ; in allusion to the foliage	169, 170
Xyrid'eæ ; from Xyris	11, 173
Xyr'is ; *xuron*, a razor ; alluding to the sharp-edged leaves of some species	173
Zale'ya ; doubtful derivation	86, 87
Zie'ria ; after J. Zier, a Polish botanist	22
Zoste'ra ; leaves resembling a belt (*zoster*)	175, 176
Zygophylle'æ ; from Zygophyllum	4, 25
Zygophyl'lum ; Gr. having yoked or paired leaves	25

ADDITIONS AND CORRECTIONS.

Page 16.—Under Cardamine eustylis, read seeds in two rows.
Page 17.—For Stenopetalum croceum, read trisectum.

Page 24.—After Eriostemon lepidotus, add:—
    Leaves small, narrow, closely revolute ... *stenophyllus*

Page 25.—After Tribulus macrocarpus, add:—
    Each fruitlet with very prominent much compressed angles, and 2 slender spines ... *Forrestii*
    Each fruitlet winged at the angle, without prickles. Shrubs.
        Glabrous; sepals woolly inside; fruitlets smooth ... *platypterus*
        Hirsute; fruitlets strongly veined ... *hirsutus*

Page 28.—After Dodonaea boronifolia, add:—
    Leaflets lanceolate, numerous, with recurved margins; rhachis dilated; lower leaves sometimes entire; broadly lanceolate ... *macrozyga*

Page 33.—For Gossypium australis, read australe.

Page 50.—After Kochia villosa, add:—
    Fruit-calyx glabrous, pale-brown, of a spongy texture, wrinkled when dry; otherwise much like *K. villosa* ... *spongiocarpa*

Page 56.—Under Urticaceae, add:—

Trema.

    Flowers polygamous in small axillary cymes; calyx-segments of male flowers induplicate-valvate in the bud; fruit a drupe, the endocarp pitted outside.
        A tall shrub, with villous branchlets; leaves ovate-lanceolate, shortly serrate, scabrous above and hirsute below ... *cannabina*

Page 66.—For Crotolaria, read Crotalaria.
    After C. medicaginea, add:—
    Leaflets 3, obovate or orbicular, very obtuse; calyx deeply lobed; standard almost acute; flowers small, few in a short raceme; ovules many; pod oblong, hairy ... *incana*

Page 74.—After Acacia scirpifolia, add:—
Phyllodia linear-subulate, 3 to 6 in. long, slightly flattened, glabrous, obscurely 1-veined on each side; peduncles 1-headed; *sepals spathulate*, not truncate; *funicle not folded* ... *juncifolia*

Page 76.—For pycynantha, read pycnantha.

Page 78.—After Acacia Kempeana, add:—
Phyllodia 5- to 9-nerved, about 4-in. long, very broad, obliquely narrowed at both ends, with a terminal gland; spikes nearly sessile; calyx 5-lobed, petals keeled ... *acradenia*

Page 82.—Under Conospermum, add:—
Leaves linear, *2 to 3 in., erect;* calyx-segments about as long as the tube, not shorter ... *Mitchelli*

Pages 83, 84.—For Hakea multistriata, read multilineata.
After H. Ednieana, add:—
Leaves terete, 4 to 6 in., simple (or dichotomously divided); flowers purple in short axillary corymbs; calyx and pedicels glabrous; fruit ovate, scarcely beaked ... *purpurea*

Page 87.—For Trianthema crystallinia, read crystallina.

Page 88.—Under Rotala, add:—
Leaves narrow, in whorls, sometimes of irregular size; capsule 3-valved; stamens 3 to 5 ... *verticillaris*

Page 94.—After Eucalyptus terminalis, add:—
Leaves opposite, ovate-cordate, sessile, rough; umbels paniculate, terminal, *rough with hispid hairs;* fruits about ½ in., or more, long, somewhat urceolate ... *setosa*

Page 96.—For Pomaderris mrytilloides, read myrtilloides.
Pages 107, 108.—For Spermacocce, read Spermacoce.
Page 122.—For Rutidosis Pumilio, read Pumilo.
Page 123.—For Podolepis Siemessenia, read Siemssenia.
Page 126.—For Helipterum Charleysae, read Charsleyae.
Page 128.—For Angianthus pussillus, read pusillus.

Page 129.—After Calocephalus platycephalus, add:—
Small erect woolly-tomentose annual; compound heads depressed-globular; phyllaries with yellow tips ... *Dittrichii*

Page 137.—For **Choripetaleae**, read **Synpetaleae**.

Page 156.—Under Spartothamnus, add:—
Stellately downy; leaves larger, flower-stalks shorter, corolla stellate-hairy outside ... *puberulus*

Page 180.—After Lipocarpha, add :—

Fuirena.

Hypogynous scales 3, flat; spikelets in paniculate clusters.
    Leaves glabrous or ciliate; flowering bracts with recurved points; hypogynous scales cordate, stalked, alternating with bristles... ... ... *glomerata*

Page 183, line 22.—For *darf*, read *dwarf*.

www.ingramcontent.com/pod-product-compliance
Lightning Source LLC
Chambersburg PA
CBHW022041230426
43672CB00008B/1036